Mathematical Expeditions

*Aristippus Philosophus Socraticus, naufragio cum ejectus ad Rhodiensiū
litus, animadvertisset Geometrica schemata descripta, exclamavisse ad
comites ita dicitur, Bene speremus, Hominum enim vestigia video.*
Vitruv. Architect. lib.6. Præf.

Mathematical Expeditions

Exploring Word Problems across the Ages

Frank J. Swetz

The Johns Hopkins University Press

BALTIMORE

© 2012 The Johns Hopkins University Press
All rights reserved. Published 2012
Printed in the United States of America on acid-free paper
2   4   6   8   9   7   5   3   1

The Johns Hopkins University Press
2715 North Charles Street
Baltimore, Maryland 21218-4363
www.press.jhu.edu

Library of Congress Cataloging-in-Publication Data

Swetz, Frank.
Mathematical expeditions : exploring word problems across the ages / Frank J. Swetz.
p.   cm.
Includes bibliographical references and index.
ISBN-13: 978-1-4214-0437-0 (hardcover : alk. paper)
ISBN-13: 978-1-4214-0438-7 (pbk. : alk. paper)
ISBN-10: 1-4214-0437-0 (hardcover : alk. paper)
ISBN-10: 1-4214-0438-9 (pbk. : alk. paper)
1. Word problems (Mathematics)—History.   2. Mathematics—History—Miscellanea.
I. Title.
QA43.S96 2012
510.76—dc23
2011021674

A catalog record for this book is available from the British Library.

Frontispiece: Frontispiece of David Gregory's *Euclidis quae supersunt omnia* (Oxford, 1703), depicting survivors from a shipwreck who are encouraged by the geometrical drawings they find on the beach. The Latin inscription reads, "We can hope for the best, for I see signs of men."

Illustration on page x: Frontispiece of Leonhard Euler's *Introduction to the Analysis of the Infinite* (1748), showing two eighteenth-century women involved in mathematical problem solving. Since women of this time period were not supposed to be involved in mathematics, the intent of this illustration is open to conjecture.

Special discounts are available for bulk purchases of this book. For more information, please contact Special Sales at 410-516-6936 or specialsales@press.jhu.edu.

The Johns Hopkins University Press uses environmentally friendly book materials, including recycled text paper that is composed of at least 30 percent post-consumer waste, whenever possible.

Contents

CONTENTS

Preface

For many years, I have worked with teachers and students to promote the inclusion of historical material into the teaching of mathematics. Such material helps to humanize mathematics—that is, associate it with its human roots by answering such questions as, Why did mathematics come into being? How was it used? Why is it important? Students need this information to help them satisfy their persistent and plaguing doubts: Why do we need this stuff? When are we going to use it? In already crowded curriculums, the introduction of more material is difficult. But I have found that an efficient, fruitful, and appealing way of incorporating historical background into mathematics teaching is through the use of actual problems posed and solved by our forebears. In talks and professional activities on this topic, I have received feedback from teachers supporting this approach.

In a previous activity book designed as a teaching resource for use in mathematics classrooms, *Learning Activities from the History of Mathematics* (Portland, ME: J. Weston Walch, 1994), I included a section on mathematical problems. These problems were received with enthusiasm by readers who had successfully employed them with their students. Later, as an editor of the e-journal *Loci*, published by the Mathematical Association of America, I conceived of and compiled the feature "Problems from Another Time," which offered a wide selection of historical problems for classroom use. Once again, teachers appreciated this effort and found this content highly motivational in classroom lessons and assignments.

Encouraged by such a reception, and of the firm opinion that such problems are indeed a worthwhile resource in teaching, I have devoted this book completely to the subject of using historical problems in mathematics teaching. The materials included are intended to be particularly suited to the classroom needs of middle school and secondary school mathematics teachers and students. University students in general mathematics classes or studying the history of mathematics will also benefit from considering these problems and their implications. The first two chapters help establish a perspective on the importance of historical problems in the learning process and in the understanding of mathematics and present strategies for the use of historical problems. Chapters 3–16 then provide a collection of approximately 500 problems. Chapter 17 presents solution techniques most likely used at the time of the problems' conception, the methods employed by the first solvers of these problems. In chapter 18 I discuss the use of historical problems and encourage the reader to seek out other such problems.

The problems have been selected for appropriate mathematical content and for the diverse social and cultural stories they tell about the uses of mathematics. Their origins range from cuneiform inscriptions on Old Babylonian clay tablets of 2000 BCE to the Egyptian Rhind Mathematical Papyrus (1650 BCE) and the Chinese mathematical handbook *The Nine Chapters* (ca. 100 CE), from the first printed European arithmetic, the *Treviso Arithmetic* (1478), to the eighteenth- and nineteenth-century *Ladies Diary* and the *Farmer's Almanac* of the American frontier. Each regional collection of problems is prefaced by some relevant comments and supported by selected illustrations.

The consideration and solution of such problems can provide a classroom introduction for a new mathematical concept or further reinforcement for a concept already learned. Each problem, in itself, also provides a brief anecdote on the needs for doing mathematics. By their context, problems also tell the reader something about the lives of people in the time of their origins. Their contents connect mathematics with society. In this respect they are open-ended and can be used for interdisciplinary teaching and a variety of classroom discussions.

Supplementing the collections of problems are brief, one- or two-page digests entitled "What Are They Doing?" These digests are intended to further enrich an understanding of historical mathematics problems by highlighting some noteworthy cultural and historic features encountered in old problems and providing insights into their solution processes. Answers to the problems can be found at the back of the book, together with a glossary of unfamiliar terms and a bibliography that includes suggestions for further reading as well as the sources cited in the text.

It is my hope that teachers will experiment with the ideas and concepts presented here; go on expeditions of learning and understanding with their classes; and eventually seek out their own collections of historic problems from which they and their students can obtain a deeper appreciation of mathematics and its human origins.

Mathematical Expeditions

1   Word Problems

Footprints from the History of Mathematics

Johannes Kepler's sketch of the orbit of Mars, published in his *Astronomia Nova* (1609). It shows the apparent retrograde motion of Mars as recorded from 1580 to 1596 according to the prevailing Ptolemaic theory. In its time, this diagram was called *panis quadragesimalis*, or the "Lenten pretzel." Using accurate observational data that he had accumulated over 15 years, Kepler refuted this theory and established an elliptical orbit for the planet.

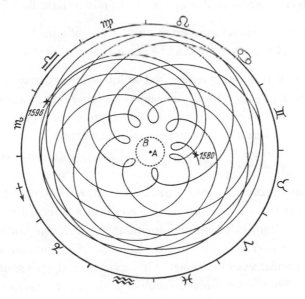

A Perspective

Historically it is interesting, and a bit telling, that some of the earliest written communications consisted of "word problems." These exercises in mathematics learning are thousands of years old and first appeared in the Tigris-Euphrates basin, ancient Mesopotamia. The thousands of clay tablets that archeologists have discovered in this region reveal the evolution of writing. The earliest tablets recovered contain imprints of clay tokens, concrete mathematical counters whose impressions designated numerical values. Gradually, this system of symbols became incorporated into a more flexible form of writing, the cuneiform script.

It appears that the initial products of this process were numerical tablets, records containing data on societal activities such as harvest amounts or taxes collected and "problem texts," that is, collections of problems for which a mathematical solution was sought or a particular problem with an answer and the solution process outlined. A dig at a temple precinct from the Sumerian city of Shuruppog unearthed the oldest known mathematics word problem, composed during the fourth millennium BCE.

A granary of barley. One man received 7 sila [of grain]. What are its men? [I.e., How many men can be given a ration?] (Robson 2007, 75)

A scribe answering this question had to know the capacity of a granary, 2400 *gur*, where 1 *gur* = 480 *sila*; the tablet provides a correct answer in sexagesimal notation. While the contents of such problems appear rather mundane now, their very existence attests to the importance of word problems.

Word or story problems are a natural extension of oral teaching, in which instead of having only a transitory verbal confrontation with a problem, one has a prolonged involvement. Eventually, the written form allows for problems to be standardized and establishes a record of important mathematical knowledge and the situations to which that knowledge is to be applied. In a sense, it is an instrument of socio-mathematical indoctrination. It specifies what mathematics is important and what situations warrant its use. Thus, word problems provide a historic testimony to the societal uses and the changing passage of mathematics over time—specifically, how mathematics was used and what its societal applications were. Word problems are the "footprints" in the history of mathematics and its teaching. Mathematically speaking, they show where we have been and the direction in which we are proceeding. Their path delineates a journey of involvement and understanding and illustrates the power of mathematics.

A good tracker can tell much from footprints. A deer hunter following his quarry can determine if the deer is moving fast, whether it has jumped, foraged for acorns on the forest floor, or been joined by companions; and he may even determine where it might be overtaken. Similarly, word problems leave a trail. They can tell their pursuer how mathematics was used, for what tasks, and reveal societal concerns and priorities.

Word problems as an extension of an oral tradition of learning supplemented that tradition and eventually replaced it by becoming a means for self-study. Before the appearance of "books," such problems served as the bare-bones collection of knowledge that had to be preserved and promoted. The appearance of word problem texts followed the rise of urbanization in southern Mesopotamia and the formation of a highly centralized state. The state was ruled by a despot and controlled through a large bureaucratic network. Most of the problem texts excavated were school texts intended for the training of scribes for state service. Mesopotamia's ascendancy initially rested upon agriculture, which in turn depended on irrigation and water conservancy. In the theory of social anthropologist Karl Wittfogel, such "hydraulic societies" share common characteristics, principal among which is a dominant bureaucracy whose task is to initiate and maintain public works projects such as the construction of dikes, canals, and grain storage facilities, and to direct land usage and tax collection (Wittfogel 1957). All of these topics are evident in the extant problem tablets. Within these collections of problems emerge a pattern of societal concerns and a chain of situations demanding mathematical consideration (see fig. 1).

Figure 1.1. Schema of problem development as based on human needs.

Food		Construction		Labor		Trade & Commerce
Planting	→	Surveying	→	Wages	→	Money
Harvesting		Area		Taxes		Partnership
Distribution		Volume		Value		Profit
Storage		Geometric		Proportion		Loss
Calendar		concepts		Social		Interest
				divisions		Measurement

Scribal problem solving follows a strict procedure and is designed to obtain a number. Numeric computation is stressed. Problem situations are frequently couched in the measurement of everyday objects and activities. Scribes are asked to find the area of fields, the lengths of canals, the amount of dirt removed from an excavation, or the number of bricks required for a structure.

> I have two fields of grain. From the first I harvest $2/3$ a bushel of grain/unit area; from the second, $1/2$ bushel/unit area. The yield of the first field exceeds the second by 50 bushels. The total area of the two fields together is 300 square units. What is the area of each field? (Van der Waerden 1983, 158)
>
> A man carried 540 bricks for a distance of 30 rods. [For this] they gave him 1 *ban* of grain. Now he carried 300 bricks and finished the job. How much grain did they give him? [1 *ban* = 10 *sila* (liters)] (Robson 2007, 115)

Despite this seemingly close association with daily events, the resulting mathematical scenarios are often unrealistic. The situation setting is merely a backdrop for the mathematics. How mathematics dominates the application is best illustrated in geometrically conceived problems, such as the following.

> A triangular piece of land [in form of right triangle] is divided among six brothers by equidistant lines constructed perpendicular to the base of the triangle. The length of the base is 390 units and the area of the triangle is 40,950 square units. What is the difference in area between adjacent plots of land? (Neugebauer and Sachs 1945, 52)
>
> A reed stands against a wall. If I go down 9 feet [from the top], the lower end slides away 27 feet. How long is the reed? How high is the wall? (Van der Waerden 1983, 57)

And there are problems undesignated in their mathematical intent—for example, "the measuring of stones":

> I found a stone but did not weigh it; after I subtracted $1/7$ and then again subtracted $1/13$ [of the remainder], I weigh it at 1 *manna*. What was the original weight of the stone? (Katz 2003, 27)

Problem collections with similar subjects and formats have been found in other "hydraulic societies" of the ancient world besides Mesopotamia: Egypt and China. One of the few extant collections of Egyptian word problems is found in the Rhind Mathematical Papyrus. This collection of 85 problems compiled in approximately 1650 BCE was prepared for scribal training. Each problem is associated with an aspect of Egyptian daily life:

Divide 100 loaves among 10 men—including a boatman, a foreman, and a doorkeeper, who received double portions. What is the share of each? (Chace 1979, 84)

How many cattle are in a herd when $\frac{2}{3}$ of $\frac{1}{3}$ of them makes 70, the number due as tribute to the owner? (Chace 1979, 102)

The theory that the great pyramids of Egypt were built by slave labor gangs has been put to rest. Archeologists now realize that they were constructed by skilled laborers, who were paid for their work in rations of grain, bread, and beer. Egyptian problem collections confirm this fact.

Perhaps the most organized, comprehensive, and influential collection of word problems from the ancient world is found in the Chinese *Jiuzhang suanshu* (ca. 100 CE). The 247 problems of this collection, together with their solution procedures and attached commentaries, served the bureaucratic needs of the Chinese empire. Its nine chapters are each devoted to specific applications of mathematics:

1. field measurement
2. processing millet and rice
3. distributions by progressions
4. short width: measurement and surveying
5. construction consultations: engineering works
6. impartial taxation: taxes and labor assignment
7. "excess and deficiency": linear equations
8. way of calculating by tabulation: systems of equations
9. right triangles: surveying

This collection became a mathematical classic, serving as a Chinese reference for over a thousand years and becoming a basic mathematical manual for Japan and Korea as well.

Westward, on the shores of the Mediterranean, Greek civilization had been developing in a series of independent city-states supported by maritime trade. The Greek empire was not a hydraulic society—one dependent on irrigation—and its intellectual characteristics and priorities differed from those of its neighbors to the east. Although mathematics was needed for social and economic uses, the Greeks established a dichotomy between applied mathematics, *logistica*, calculation, and theoretical mathematics, *arithmetica*, aspects of numbers and shapes worthy of philosophical consideration. *Logistica,* the "less worthy" mathematics, was undertaken by slaves, craftsmen, and merchants. No records of their mathematical problems remain. In the hydraulic societies, mathematical problems had become an end in themselves and resulted in specific numerical answers, whereas for the Greeks problems were a beginning from which theories evolved. The earliest surviving collection of Greek word problems is found in the Greek Anthology and was probably assembled by the grammarian Metrodorus (ca. 500 CE). The 46 arithmetical problems are stated as riddles and appear to be compiled for intellectual challenge.

Word problem collections would play a prominent role in the teaching of commercial mathematics and the introduction of the Hindu Arabic numerals and their computational algorithms into medieval and early Renaissance Europe. As textual forms became more elaborate, the number of problems introduced into a discourse decreased; however, problem scenarios would be repeated, and a standardization of word problems emerged. Word problems now supplemented and reinforced more textual instructional forms. They appeared in arithmetic books, but their emphasis still varied according to existing social conditions and the intent of the author. In sixteenth-century England, writers of mathematics texts such as Robert Recorde (1510–58) and Humphrey Baker (d. 1587) presented problems that appealed to craftsmen and tradesmen. Recorde also posed problems related to warfare and military affairs, as did several other authors of arithmetic texts at this time—for example, Christoff Rudolff, Niccolo Tartaglia, and Thomas Digges.

With the popularization of mathematics in the eighteenth and nineteenth centuries came the rise of periodicals—newspapers, almanacs, and journals—that published collections of word problems for the mathematical edification and education of their readers. One influential such journal was the *Ladies Diary*, published in England from 1704 to 1841. Challenge problems would be offered to readers and their correct solutions published in later editions. When the first mathematical periodical, the *Mathematical Correspondent*, was published in the United States in 1804, it emulated the *Diary* in its exposition and use of problems and even offered a cash prize for the best problem solutions.

Problem Forms and Presentations

The earliest written problems were simple statements followed by a question, capturing the words of the mathematics master or instructor for permanence and reuse. They tested basic comprehension and operational methods: find a sum, divide the amount, and so on. But gradually they became more demanding, requiring analysis of data and synthesis of solution methods. Their social relevance and psychological appeal were increased by reference to daily activities. Such allusions to real life could result in problems with realistic data producing practical solutions or in problems with pseudo-realistic settings where impractical answers resulted. This latter class of problems was usually conceived to demonstrate mathematical concepts rather than to be utilitarian. Subclasses of realistic problems that emerged in Europe by the sixteenth century were scientifically based queries whose answers entailed knowledge both of mathematics and of the new sciences.

A third class of word problems were developed as recreational, intellectual challenges. In most early problem collections, such "riddles" or "puzzles" were interspersed among the practical mathematics problems; they stood as an assertion that mathematics was also a pure intellectual activity transcending the realm of daily activities. Perhaps the simplest and most enduring of these recreational problems are ones in the form "guess my number," as first found in the 3,000-year-old Rhind Mathematical Papyrus:

> What quantity whose whole and seventh added together gives 19? (Chace 1979, 66)

And in the early United States:

> There are two numbers whose sum is equal to the difference of their squares; and if the sum of the squares of the two numbers be subtracted from the square of their sums, the remainder will be 60. What are the two numbers? (Watson 1777)

A fascination with geometric progressions appears in many early recreational problems and spans various cultures. The Rhind papyrus introduces the proverbial "seven cats":

> [There are] seven houses; in each 7 cats; each cat kills 7 rats; each rat would have eaten 7 ears of spelt; each ear of spelt will produce 7 *hekat*. What is the total of them all? How much *hekat* of grain is thereby saved? (Chace 1979, 112)

Two thousand years later, an independent Chinese version in *Master Sun's Mathematical Manual* (ca. 400 CE) replaced the Egyptians' culturally preferred number seven by the Chinese special number nine:

> Now there is sighted 9 embankments; each embankment has 9 trees; each tree has 9 branches; each branch holds 9 nests; each nest has 9 birds; each bird has 9 young; each young has 9 feathers and each feather has 9 colors. Find the quantity of each. (Lam and Ang 1992, 181)

When the British monk Alcuin of York became educational advisor to the emperor Charlemagne in 781, he compiled a collection of mathematical word problems for the training of court pages. His *Propositiones ad acuendos juvenes* was a collation of 56 riddle-type problems, the first Latin collection of such problems. It too considered questions of progressions:

> A ladder has 100 steps. On the first step sits 1 pigeon; on the second 2; on the third 3 and so on up to the hundredth. How many pigeons are there in all? (Hadley and Singmaster 1992, 121)

When Leonardo of Pisa published his influential *Liber Abaci* in 1202, he included several of Alcuin's problems among his collection of reader exercises. Once established, problem types and scenarios provided templates upon which future problems would be composed for other generations.

Mathematical Content

Several years ago I conducted an experiment with an audience of 30 experienced secondary school teachers. They were participants in a workshop on the history of mathematics that I was giving. As a preliminary test of their knowledge, I gave them five historical word problems disguised so that the problems' cultural and temporal origins remained unknown. In doing the problems and examining the mathematical concept employed in each, they

were asked to date the origin of the exercise within several given chrono-
logical categories (ancient Egypt, Renaissance Europe, etc.). If we consider
70% correct a passing grade, 28 of the teachers failed the task. Quite simply,
they did not know when the mathematics they had been teaching for years
originated. Since the duration of a mathematical concept over time attests
to its importance, these teachers had a limited appreciation for the relevance
of the mathematical concepts included in the test.

The mathematical content in word problems compiled a millennium
before the Christian era reveals that the calculators of the time could perform all the basic operations known today with a high degree of accuracy.
They could extract square and cube roots to several decimal places; they
knew correct formulae for most common areas and volumes. They worked
with linear and quadratic equations, understood the concepts of arithmetic
and geometric progressions, approximated a value of π to a workable accuracy, and knew the mathematical relationship popularly known as
the Pythagorean theorem. Pythagoras flourished in the fifth century BCE,
yet problems whose required solution demonstrates knowledge of the relationship that bears his name are found in many problem collections that
predate him by centuries. For example, there are these Old Babylonian
(2000–1600 BCE) problems:

> A beam of length $\frac{1}{2}$ [stands against a wall]. The upper end has slipped
> down a distance $\frac{1}{10}$. How far did the lower end move? (Van der
> Waerden 1983, 59)
>
> [Given] a gate, the height is $\frac{1}{2}$ rod, 2 cubits, the breadth 2 cubits.
> What is the diagonal? (Robson 2007, 140)

The variety and scope of the 24 right triangle problems given in the
ninth chapter of the *Jiuzhang* indicate that Chinese surveyors and mathematical scribes were well aware of the Pythagorean theorem at an early
date. In particular, two of the problems in this collection, the "broken bamboo" and the "reed in the pond," have since appeared in different guises in
several cultures:

> A bamboo shoot 10 *chi* tall has a break near the top. The configura-
> tion of the main shoot and its broken portion forms a triangle. The
> top touches the ground 3 *chi* from the stem. What is the length of the
> stem left standing? (Swetz and Kao 1977, 44)

In the center of a square pond whose side is 10 *chi* grows a reed whose tip reaches 1 *chi* above the water level. If we pull the reed towards the bank, its top becomes even with the surface of the water. What are the depth of the pond and the length of the plant? (Swetz and Kao 1977, 30)

The bamboo problem appears again in the ninth-century Sanskrit mathematical classic *Ganita-Sara-Sangraha* by Mahavira and still later in Filippo Calandri's *Arithmetic*, published in Florence in 1491 (fig. 1.2). In a more picturesque Hindu version of the reed problem published by

Figure 1.2. (a) Bamboo problem from chapter 9 of the *Jiuzhang suanshu* (China, ca. 100). (b) The broken tree problem as given in Filippo Calandri's *Arithmetic*, published in Florence in 1491. This was the first illustrated, printed arithmetic book to appear in Europe.

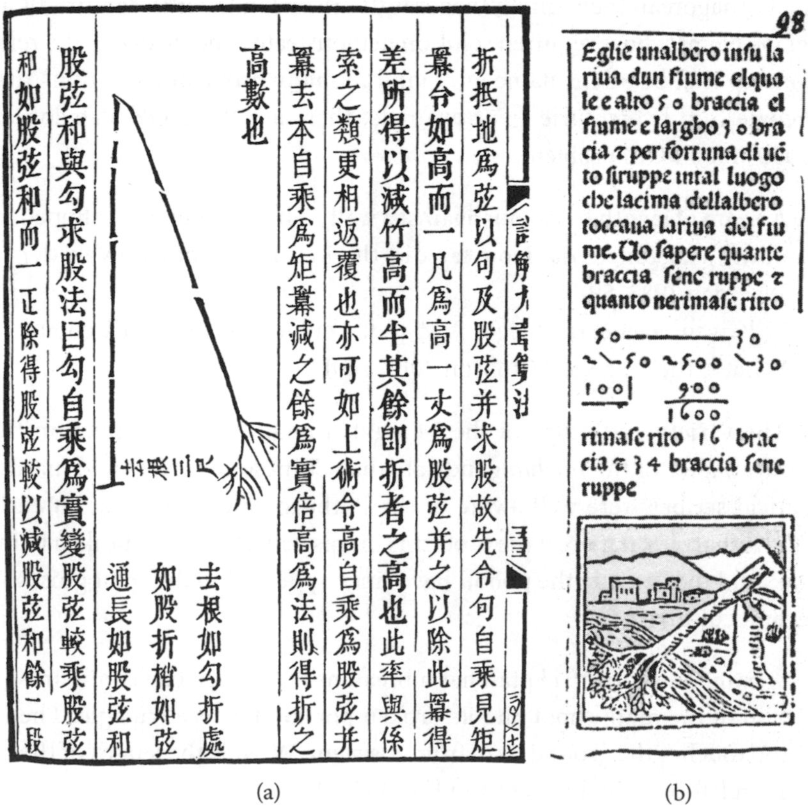

(a)

(b)

Bhaskara (1114–ca. 1182), the reed becomes a lotus, and red geese occupy the pond (Colebrooke 1817, 66).

Further inspection of the problems in the eighth chapter of the *Jiuzhang* reveals that Chinese computers of this early period could solve systems of linear equations by Gaussian elimination, a method attributed to C. F. Gauss (1777–1855) and worked easily with positive and negative numbers, a feat previously first credited to Hindu mathematicians of the seventh century.

There are some instances where the first appearance of a mathematical technique is given in a problem. For example, the first known example of the Chinese "remainder theorem" appears in a problem from *Master Sun's Mathematical Manual* (400 CE):

> Now there are an unknown number of things. If counted by threes there is a remainder of two; if counted by fives there is a remainder of three; and if counted by sevens there is a remainder of two. Find the number of things. (Lam and Ang 1992, 104)

From the thirteenth century onwards, Chinese word problems demonstrated techniques for extracting numerical roots for higher-degree equations. This facility was not common in Europe until the appearance of the Ruffini-Horner method in 1819. A typical problem of this type is the following:

> There is a tree 135 *bu* from the southern gate [of a walled city]. The tree can be seen if one walks 15 *bu* from the northern gate and then 208 *bu* in the eastward direction. Find the diameter of the walled city. (Li Zhi 1248)

Using modern notation and allowing the radius of the city to be represented by x, the conditions result in the equation

$$4x^4 + 600x^3 + 22500x^2 - 11681280x - 788486400 = 0,$$

and x is correctly found to be 120 *bu* (Libbrecht 1973, 134).

Word problems posed as simple statements intended to foster mathematical thinking have often served as a seed for mathematical research. Consider the "pursuit problem," where one creature, person, or animal is pursuing another. The first known appearance of this problem once again comes from the Chinese *Jiuzhang*, chapter 6, problem 14:

A hare runs 100 *bu* ahead of a dog. The dog pursuing at 250 *bu* is 30 *bu* short. Tell me in how many *bu* will the dog catch up with the hare? (Shen, Crossley, and Lun 1999, 330)

Alcuin's European version is this:

A dog chasing a rabbit, which has a start of 150 feet, jumps 9 feet every time the rabbit jumps 7. In how many leaps does the dog overtake the rabbit? (Hadley and Singmaster 1992, 115)

By the early Middle Ages, the European alteration by Abraham ben Ezra (ca. 1140) had replaced the animals with travelers, an example which was probably more relevant to the contemporary scene. The situation has been personalized by the insertion of particular names:

Reuben sets out from his city on the morning of the first day of the new moon to go to meet his brother Simon in Simon's town. On the same day, Simon also leaves his town to go see Reuben in his city. The distance between the two places is 100 miles. We ask "when will they meet?" (Sanford 1927, 72)

Still later, the situation was generalized further by the insertion of couriers. "Courier problems" with various routes taken remained popular among German and Italian writers of the sixteenth century. In 1732, the French mathematician Pierre Bouguer proposed a pursuit problem before the French Academy in which he envisioned a merchant ship and a pirate ship. The merchant flees the pirate on a course perpendicular to that of the pirate who pursues. Bouguer sought the *courbe de poursuits*, or "curve of pursuit." This situation initiated a topic of mathematical research that exists until the present day: interception and pursuit analysis.

Another simple problem that has spawned some interesting variations is the "river crossing" problem. Alcuin of York first gave it in his *Propositiones* as follows:

Three friends each with a sister needed to cross a river. Each of them coveted the sister of another. At the river they found a small boat in which only two of them could cross at once. How could they cross the river without any of the women being defiled by the men? (Hadley and Singmaster 1992, 111)

Three other problems in Alcuin's collection involve the same situation with different subjects: wolf, goat, and cabbage; overweight people; and hedgehogs. Perhaps the licentious implications prompted continual interest in this problem. Luca Pacioli, a Renaissance mathematics teacher, asserted that four or five couples will require a three-person boat. Mathematician Niccolo Tartaglia in 1556 claimed that four couples could cross in a two-person boat, challenging Pacioli. Finally, Claude-Gaspard Bachet de Méziriac refuted Tartaglia in his own collection of recreational problems, *Problémes plaisants et delectables*, in 1624. Interest in this problem was resurrected in the nineteenth century by the French mathematician François Édouard Anatole Lucas. In 1879, one of Lucas's students, Cadet de Fontenay, introduced an island into the situation that allowed four couples to complete the transit in 24 crossings (Pressman and Singmaster 1989). Since that time, Rouse Ball has concluded that $6n - 7$ crossings are required for n couples (Ball 1987). The cultural variants of this particular problem are also interesting and will be examined later.

Where the Footprints Lead:
Societal and Cultural Relevance

Mathematical historian D. E. Smith, in his early-twentieth-century examination of problem collections (Smith 1918), was impressed by the amount of societal and cultural information conveyed by word problems. Word problems, by their content and emphasis, are vehicles for societal indoctrination either explicitly or implicitly. The mathematical tracker can discern the subject's activities during a journey. Much factual and historical information can be revealed from word problems.

In ancient Mesopotamia, the transition from the fourth millennium to the Old Babylonian period is marked by a lessening of despotic bureaucratic rule: city-states express their independence; collective agriculture is replaced by smaller holdings; royal workshops are replaced by private handicraft; royal traders become independent merchants; and individuals are allowed to use identification seals, a practice formerly reserved for royalty. There is a shift of importance from the state and its apparatus to individuals. A new self-confidence is evident in scribal writing. Written mathematical problems become more creative, algebraic in their conception, and more personally directed as mental challenges rather than mere exercises.

By contrast, in Chinese problem collections a striking similarity and rigid form of problem posing and solving is maintained over centuries. In the Confucian tradition, classics written by masters were to be revered, copied, and perhaps respectfully commented upon. The standard was also followed in mathematics instruction, where all texts imitated a canon of problem forms, or recipes. Nineteenth-century problem collections still mirrored the situation depicted in the first-century *Jiuzhang*. This dogmatic adherence to set problem forms and standards stifled creativity and limited mathematical advances in the Chinese empire. Because the mathematical question "What if?" did not appear, neither did theoretical mathematics (Swetz 1996).

While the Japanese initially emulated their Chinese neighbors and adopted China's mathematical classics and problem collections for the instruction of their scholars and bureaucrats, by the Edo period (1603–1868) they began evolving their own problem forms. During this period Japan retreated into isolationism, removing itself from feared Western encroachment. It was a time for cultural introspection and renewed reverence for traditional customs. In 1627 the mathematician Yoshida Kōyū published *Jinkoki* (Treatise on eternal mathematical truth), in which he set a new mathematical standard by concluding his work with a list of unsolved problems. These were taken up by readers, who, in turn, published their solutions and offered further challenge problems. This soon evolved into a popular wave of problem solving based mainly on the solution of complex and fanciful conceived geometric problems involving circular properties and dealing with finding the lengths of chords, arcs, and the like. Common people, such as farmers, now openly became involved in problem solving; and as a gesture of thanksgiving, and perhaps an expression of bravado for solving a problem, they inscribed their problem and solution on a wooden plaque and hung it in the local temple or Shinto shrine. Often these tablet problems, *sangaku,* bore the challenge, "See if you can prove this!"

Historically, such a movement in problem posing and solving appears unique to the Japanese. One of the Kōyū problems is as follows:

There is a log of precious wood 18 feet long whose bases are 5 feet and 2½ feet in circumference. Into what lengths should this log be cut to trisect its volume? (Cooke 1997, 248)

And a *sangaku* in modern notation:

The centers of a loop of n equal circles of radius r are the vertices of an n-gon. Let S_i be the sum of the areas inside the n-gon, S_o, the sum of the areas outside. Show that $S_i - S_o = 2\ r$. (Fukagawa and Rigby 2002, 27)

The rise of European mercantile capitalism in the late Middle Ages saw an accompanying renewed interest in numerical computation and problem solving. Between the twelfth and the fifteenth centuries came a wave of manuscripts and eventually books that promoted a use of Hindu Arabic numerals and their associated algorithms and commercial problem solving. Problems now accompanied more theoretical explanations, but they still supplied the majority of practical instruction. When the first printed arithmetic book, the *Treviso Arithmetic,* appeared in Italy in 1478, its 123 pages of text included 62 problems. A variety of business issues were covered by such problems: computation of interest, determination of profit and loss, insurance, barter, the mathematics of partnership, *tret* and *tare*, payment of taxes and custom duties, transport times and cost, allegation or mixing problems, and many more topics of concern (Swetz 1987).

The situational descriptions provided in these problems reveal much about the daily life of the times: oxen were used for plowing; important centers of European trade were Venice, Lyons, London, and Antwerp; the commodities most often traded were cloth, wool, brass, and rice. In Sienna, the rental price of a house was 30 lire a year, while in 1640 Florence, the price was 300 lire. Beef selling for 1 *grosso* for 3 pounds in fourteenth-century Italy was intended for the wealthy, as were spices such as pepper, ginger, and sugar. Bread in sixteenth-century Italy cost a half penny for 20 ounces. The existence of a social class structure is obvious. The relative quality of some products can be judged from the information given: Spanish linen sold for 94 to 120 ducats per hundredweight, while Italian linen fetched 355 ducats for the same amount.

A predominance of monetary exchange problems as well as mixing (allegation) problems regarding metals indicates the confused state of minting and monetary standards. This fact is further confirmed by considerations of barter, indicating the unreliability of the money supply. "Tret" and "tare" are antiquated topics but important ones for the merchants of Renaissance Europe. Tret is the weight allowance for the heavy containers—barrels, crates, and the like—used to package goods; tare is the wastage due to the hardships of transport. Travel times and the imposition of multiple custom duties also testify to the difficulty of moving commodities. Even hotel life is revealed in a German problem from 1561 which describes a *Gausthause* with eight rooms, each room with 12 beds and each bed sleeping three guests. The reader is asked to compute the payment due.

Frequent reference to the wool industry, a primary source of European income at this time is found in many books:

> A man bought a number of bales of wool in London, each bale weighing 200 pounds, English measure, and each bale cost him 24 florin. He sent the wool to Florence and paid carriage, duties, and other expenses amounting to 10 florin. He wishes to sell the wool in Florence at such a price as to make 20% on his investment. How much should he charge a hundredweight of 100 London pounds, which are equivalent to 133 Florentine pounds? (Ghaligai 1521, fol. 31v)

Changing social, political, and economic needs saw the situational content of word problems change accordingly. The rise of animal husbandry in Europe was accompanied by the appearance of "pasturage" problems:

> Two men rent a pasture for 100 *livres,* on the understanding that two cows are to be counted as equivalent to three sheep. The first puts in 60 cows and 85 sheep; the second 80 cows and 100 sheep. How much should each pay? (Trenchant 1566, 178)

The early Christian church forbade the charging of interest in financial transactions, and the policy was affirmed by the church's Council of Vienne in 1311. However, commerce demanded the lending of money. With the church's approval, Jewish money lenders and bankers filled this void, giving loans and charging interest. Compound interest, an unpleasant prospect to labor under, was often associated with Jewish money lenders:

> A Jew lends a man 20 florins for four years and every half year he reckons the interest on his capital. I ask how much the 20 florins will amount to in four years if a florin earns 2 d. [pennies] a week? Find the interest on the interest. (Riese 1522, fol. Gv)

After the Lateran Council of 1515 allowed Christians to charge interest, attention to interest computation in problems increased.

By the sixteenth century, the needs of European warfare began to find their way into mathematical word problems. For example, in order to blunt cavalry charges against standing infantry, in the fifteenth century the Swiss devised the use of square phalanx formations of pikemen and halberdiers. Formed into a tight square formation and with their weapons pointed outward, this defensive formation was called a "hedgehog." The mathematics of square troop arrangements became a subject of problems.

There is a capitian, whiche hath a greate armie & would gladly marshall them into a square battaile, as large as might be. Wherefore in his first proofe of a square forme, he had remaining 284 too many. And prouying again by putting 1 moare in the fronte, he founde wante of 25 men. How many soldiers had he as you guesse? (Recorde 1557, fol. G)

Although the introduction of artillery in warfare made such formations obsolete, this problem continued in arithmetic books until the nineteenth century.

An interesting trail to follow is that of an appealing word problem through its variant forms over a period of time. A problem that has endured in almost all cultures is the "cistern problem." Probably originating in the Mediterranean world of ancient Greece or Rome, the problem initially concerned water flowing in a fountain. A version given in the Greek Anthology goes as follows:

I am a brazen lion; my spouts are my two eyes, my mouth and the flat of my right foot. My right eye fills a jar in two days, my left eye in three, and my foot in four. My mouth is capable of filling it in six hours. Tell me how long all four together will take to fill it? (Page et al. 1916, 31)

In sixteenth-century agricultural Europe, milling became the subject:

A man wishes to have 500 *rubii* of grain ground. He goes to a mill that has five stones. The first of these grinds 7 *rubii* of grain in an hour, the second grinds 5, the third 4, the fourth 3 and the fifth 1. In how long a time will the grain be ground and how much is done by each stone? (Clavius 1583, 191)

Later, the same mathematical situation is depicted in regard to trade: sails on a ship (Borghi 1484), companions drinking wine (Buteo 1559), animals devouring a sheep (Calandri 1491), and in perhaps its most recurrent form, concerning labor:

> If two men or three boys can plow an acre in ⅙ of a day, how long will it require three men and two boys to plow it? (Brooks 1873, 191)

One Victorian version given in J. H. Smith's *Treatise on Arithmetic* (1880) reflects on a miner's working conditions:

> If 5 pumps each having a length of stroke of 3 feet working 15 hours a day for 5 days, empty water out of a mine; how many pumps with a length of stroke 2½ feet, working 10 hours a day for 12 days, will be required to empty the same mine; the strokes of the former set of pumps being performed 4 times as fast as the other? (p. 120)

Another problem that has left a long and winding trail is Alcuin's river crossing situation, mentioned above. Moving beyond the Catholic Church's moral concern with the innocence of young virgins, the problem takes on social class distinctions. A 1624 version has three masters and their valets crossing the river, but each master hates the other's valet and will do him harm if given the opportunity; masters and valets must be separated. In 1881, *Cassell's Book of Indoor Amusements* depicts the situation with violent servants who will rob any outnumbered master. Ten years later, at the height of British imperialism and the carrying out of the theory of the white man's burden, the problem becomes one of missionaries and cannibals where the missionaries must keep from being eaten by their traveling companions.

Most problems considered in early American arithmetics reflected on the nation's emergence as a trading power. Such problems provide a wealth of information on the mercantile conditions of the time.

> Shipped for the West Indies 223 *quintals* of fish, at 155.6 d. per *quintal*; 37000 feet of boards at $8⅓ per 1000; 12000 shingles at ½ *guines* per 1000; 19000 hoops at $1½ per 1000, and 53 half *joes* [Portuguese coins]; and in return, I have 3000 gallons of rum at 1s. 3d. per gallon; 2700 gallons of molasses, at $5½ d. per gallon; 1500 pounds of coffee at 8½ d, per pound and 19 cwt of sugar at 12 s. 3 d. per cwt and my

charges on the voyage were £37 12s, pray, did I gain or lose, and how much by the voyage? (Pike 1788, 133)

A later problem appearing in the 1814 issue of *The Analyst* reflects on the human cost of the plantation system of the West Indies:

If out of a cargo of 600 slaves 200 die during a passage of 6 weeks from Africa to the West Indies, how long must the passage be that one half the cargo may perish? Supposing the degree of mortality to be the same throughout the passage, that is, the number of deaths at any time to be proportional to the living at the same time. (Douglas 1814, 21)

When, in 1776, England's 13 American colonies broke away and became the United States of America, each colony was an independent civic and political entity unto itself. This independence was reflected in the monetary systems the former colonies employed. All were different. As a commercial nation, the problems of monetary exchange, both foreign and domestic, became a major concern for the fledgling United States. This issue was evident in the mathematical word problems of this period.

A different kind of track is provided by problems with a political or social agenda. Problems can reveal societal divisions: favorite groups can be designated and viewed in a good light, while undesirables can also be recognized. One of the most curious of such designations is the "Josephus problem," named after the first-century Jewish historian Flavius Josephus, who found himself trapped in a cave, together with 40 colleagues, by the Roman army. Facing imminent death, the group chose suicide. Josephus arranged himself and his companions in a circle where from a certain point every third man would be killed until all were eliminated. Josephus chose his position so that he became the last man standing and so survived. A popular form of this problem comes from the tenth century and involved Turks and Christians.

A sinking ship must cast off passengers to survive. There are 15 Christians and 15 Turks aboard. The captain, himself a Christian, arranges the passengers in a circle where every ninth person will be thrown overboard. How should he make the arrangement so that the Christians survive? (Smith 1958, 541)

An eighteenth-century Japanese version has a stepmother arranging her children, both stepchildren and her own offspring, in a circle upon which a selection scheme will be applied to disinherit some children. She wishes to have her children benefit from the process but miscounts and all are disinherited. A problem with a real moral!

Another problem that has lent itself to discriminating situations is that of the "hundred fowls." Originating in China in about the fifth century, the problem initially involved chickens, ducks, and sparrows; thus, the "fowls" designation. Its solution results in a linear indeterminant equation for which a practical value is required. A twelfth-century Islamic version places Christians and Jews at a disadvantage:

> A Turkish bath has 30 visitors in a day. The fee for Jews is 3 *dirhams*, for Christians 2 *dirhams* and for Muslims ½ *dirham*. Thirty *dirhams* were earned by the bath. How many Christians, Jews and Muslims attended? (Rebstock 2007, 11)

An Italian problem celebrates the expulsion of the French from the Italian territories in 1554 by reporting:

> The King of France entered into a battle and was defeated in such a way that ¼ of his soldiers were killed, ⅔ wounded, 1000 were taken prisoner and 6000 left on the field. I want to know how many soldiers he had before he was defeated? (Swetz 1994)

More blatant propaganda can be found in modern word problem collections. In the early twentieth century the United States was attempting to instill democratic principles in the Philippines and remove Spanish colonial traditions. Special textual materials were produced to help introduce these new ideals to the subject population. Arithmetic problems focused on the evils of the existing tenant land-holding situation.

> Pedro is a tenant on Mr. Santos' farm. He has rented 4 hectares of rice land. After the cutting is paid for, Mr. Santos is to have for the use of the land one half of what rice is left and Pedro will take the other half for himself. If 45 *cavans* grow on each hectare, and one sixth is given for cutting, how many *cavans* will the cutters get? How much will be left? What will be Mr. Santos' share? What will be Pedro's share? (Bonsall 1905, 113)

The problem series goes on to point out that Pedro is in debt to Mr. Santos for seed rice and also for products purchased at Mr. Santos's store. After a series of calculations, the student-reader finds that Pedro comes out of this situation even deeper in debt.

Landlord exploitation was also a frequent topic in Chinese Communist textbooks:

> In the old society, there was a starving family who had to borrow 5 *dou* [200 pounds] of corn from the landlord. The family repaid the landlord three years later. The greedy landlord demanded 50% interest compounded annually. How much corn did the landlord demand at the end of the third year? (*New York Times* 1969)

The cost of "undesirables" in Reichsmarks was considered in Nazi textbooks used in Germany in 1941:

> Every day, the state spends RM 6 on one cripple; RM $4\frac{1}{2}$ on one mentally ill person; RM $5\frac{1}{2}$ on one deaf and dumb person; RM $5^{3}/_{5}$ on one feeble-minded person; RM $3\frac{1}{2}$ on one alcoholic; RM $4^{4}/_{5}$ on one pupil in care; RM $2^{1}/_{10}$ on one pupil at a special school; and RM $^{9}/_{20}$ for one pupil at a normal school.

There then followed a series of problems emphasizing the cost to the state for "inferiors," such as the following:

> Calculate the expenditure of the state for one pupil in a special school and one pupil in an ordinary school over eight years and state the amount of higher cost engendered by the special school pupil. (Pine 1997, 27)

Pedagogical Implications

Word problems are composed to teach mathematics. For thousands of years they were the primary means for such instruction. Their compilation required an investment of time and thought. The process of devising problems entailed selecting the mathematic principles to be communicated and embedding them in a situation that provided motivation—societal, intellectual, or both—for the reader to solve the problem. Indeed, many early problem collections were prefaced by motivational comments. For example, the author of the Rhind papyrus promises its reader that the text provides "insights into all that exists" and "knowledge of powerful secrets" (Chace 1979, 27). Sun Zi, in the introduction to his arithmetic classic,

Master Sun's Mathematical Manual (ca. 400), assures his reader that "mathematics governs the length and the breadth of the heavens and earth; affects all creatures" (Lam and Ang 1992, 151). Such admonitions as to the power and usefulness of mathematics were continued by Robert Recorde in his sixteenth-century efforts to popularize mathematics in England and became a standard feature in early American mathematics books.

Further motivation is then built into the problem sequence. Problems develop from simple statements issuing a challenge, "Find me a number . . .," to directed dialogues stressing social and economic relevance, to extended story situations that employ "human interest" to draw the reader closer into the drama:

> The burgomaster and council of the city of Oppenheym employed a learned writing master for the city, telling him if he would serve them faithfully for a year they would give him 100 guilders, a horse, and a suit of clothes. The school master taught no longer than three months when he was obliged to take leave of the council and, for his services of three months, they gave him the horse and the clothes and said, "Now take the horse and the outfit of clothes and go on your way." The writing master received them gladly and went happily on his way. The question is "What was the value of the horse and the clothes, since they served as three month's reckoning?" (Köbel 1514, fol. 78r)

Hindu problems were written in verse and referred to fantasy situations, appealing to the reader's sense of aesthetics and imagination and helping in memorization of the problems.

Pedagogical ordering is evident in many collections of problems, which progress from simple problems to more complex; from single procedural solutions to those requiring multiple computing techniques; from concrete-based problems to the abstract. A learner's confidence is established before the learner ventures into higher levels of involvement. An early visual example of this principle is evident in a British Museum cuneiform tablet, BM 15285, from the Old Babylonian period that presents a series of geometric configurations. As the scribal students progressed through the series, they encountered more intricate geometrical situations to unravel (fig. 1.3). The Chinese *Jiuzhang* clearly indicates a purposeful pedagogical ordering in both its chapters and its problem sequencing.

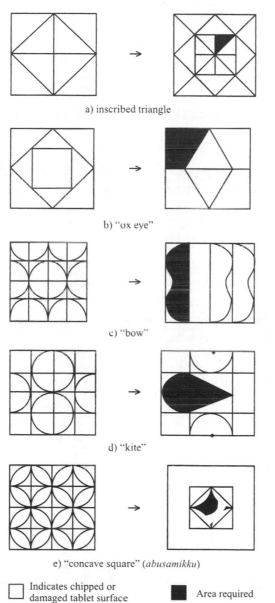

a) inscribed triangle

b) "ox eye"

c) "bow"

d) "kite"

e) "concave square" (*abusamikku*)

☐ Indicates chipped or damaged tablet surface   ■ Area required

Figure 1.3. Some exercises from British Museum cuneiform tablet 15285. This tablet is believed to have originally contained about 40 such exercises requiring scribal students to find the areas of selected regions within the given squares.

Up through the nineteenth century, word problems were carefully chosen for their impact and used to supplement instruction. They were a major part of the instruction. As the design of mathematics textbooks changed, with a stress on more theoretical aspects of mathematics, a reliance on the instructional power of word problems decreased. Word problems no longer make a statement as to the importance and power of mathematics. They have often been reduced to the status of "busy work" or employed as a means of punishment. This is a situation that demands rectification!

Conclusion

The mathematical footprints we have been following from the late Bronze Age to the present have led us on a long trail. At the start, the trail was quite narrow, but it soon broadened, became well traveled, and began to branch in different directions. Each side path—word problems revealing economic conditions; illustrating trade and commerce; reflecting on contemporary events, scientific advancement, social movements, and warfare—deserves exploration. Word problems remain a valuable resource for teaching mathematics, but also, an examination of their history can supply an understanding of the development of mathematical ideas, their priorities, and their interrelationship with the real world.

2 Problems, Problems

A Resource for Teaching

The Greek mathematician Archimedes (287–212 BCE) was considered one of the greatest applied mathematicians of all time. An avid problem solver, he was employed by King Hieron of Syracuse. During the siege of Syracuse by the Romans, he designed war machines that kept the Roman at bay for three years. Legend has it that when the Romans finally broke the siege in 212 BCE, Archimedes, deeply involved in solving a problem, ignored the command of a Roman soldier and was killed. This engraving by the nineteenth-century artist Gustave Courtois depicts Archimedes engrossed in his final problem.

Since this discussion is about problems, let us begin with a problem:

> A circular piece of land 100 measures in diameter is to be divided among three persons so that they shall receive 2900, 2500, and 2500 square measures, respectively. Find the lengths of the resulting chords and the altitudes of the segments. (Cooke 1997, 2480)

The situation appears straightforward and reasonable: the division of a piece of land into three parts. A diagram is probably desirable. But still, this problem requires quite a bit of thought. It is a good problem to explore. If it is given to a secondary school class, three questions will usually arise: "What is the answer?" "How do you do it?" and "Where did this problem come from?" Many times, it is the response to the third query that will result in further discussion and prompt still another question: "How did they do it?" This problem about a circular piece of land was posed by the Japanese mathematician Yoshida Kōyū in 1627 and was typical of the topics Asian mathematicians were involved with at this period of time: areas of circular segments. The problem is from another era and another place. Its remoteness, both temporally and culturally, from a contemporary classroom adds a mystique and intrigue to the problem and provides another fertile dimension of learning: seventeenth-century Japanese mathematicians were doing some interesting mathematics.

Mathematics teachers are always seeking "good" problems for use in their classes. The history of mathematics abounds with such problems. Since earliest times, whether written in clay, on papyrus scrolls, or on bamboo strips, mathematical word problems have served as a primary means to communicate the uses and techniques of mathematics. They remain a readily available source for classroom enrichment.

Reasons for Using Historical Problems

In the mathematics education community it is generally felt that the inclusion of mathematics history into the teaching of the subject does much to enrich the experience. History often provides answers to the whys and hows so often left out of regular classroom instruction. It answers the persistent student question, "Why are we studying this?" History also illustrates the continuity of mathematical involvement. Thousands of years ago, people were trying to solve problems similar to the ones we solve today. They were doing mathematics! This is often a startling revelation for students, but it is also a reassuring one in the sense that they realize they are part of an ongoing process.

But how can history be included easily into classroom instruction? Historical problems supply a means to satisfy this need. They are numerous and readily available. Most mathematical topics are covered in such problems, and their features make them attractive for classroom use:

- Although teachers resist more direct and extensive historical intrusions, they readily accept and use problems. Problems are nonthreatening; they are not something extra in the mathematics curriculum.
- The mathematical context of problems is relevant to instructional needs.
- The settings, historical milieu, and situational encounters of problems provide an intrigue and added motivation for students.
- Historical problems promote an appreciation of diversity, both mathematical and cultural, and supply a flexible background for interdisciplinary explorations connecting mathematics with history, culture, and economics.

Here are some historical problems that could satisfy such objectives:

Having been given the perimeter and perpendicular of a right angled triangle, it is required to find the triangle. (Isaac Newton, 1728)

A leech invited a slug for lunch a *leuca* away. But he could only crawl an inch a day. How long will it take to get his meal? [1 *leuca* = 1500 paces; 1 pace = 5 feet] (Alcuin of York, 800)

Find two numbers with sum 20 and when squared their sum is 208. (Diophantus, ca. 250 CE)

A square-walled city of unknown dimensions has four gates, one at the center of each side. A tree stands 20 *bu* from the north gate. One must walk 14 *bu* southward from the south gate and then turn west 1775 *bu* before he can see the tree. What are the dimensions of the city? (China, 200 CE)

Strategies for Incorporation

Although teachers will accept collections of historical problems for use in the classroom, they may require some supporting suggestions as to just how these problems may be efficiently employed. Of course, just as there are varied ways of teaching mathematics depending on class needs and ability, there are also varied ways of employing historical problems in the mathematics classroom. Consideration of a historic situation or problem can introduce a mathematical concept: the scaling of a wall using a ladder lends itself to the conceptualization of a base-height-hypotenuse relationship for a right triangle, the Pythagorean theorem. Indeed, right triangle relationships have been a part of mathematics learning for thousands of years, long predating Pythagoras. A variety of such problems are given below:

> A reed stands against a wall. If it moves down 9 feet [at the top], the [lower] end slides away 27 feet. How long is the reed? How high is the wall? (Babylonia, 1800–1600 BCE)
>
> An erect [vertical] pole of 30 feet has its base moved out 18 feet. Determine the new height and the distance the top of the pole is lowered. (Egypt, 300 BCE)
>
> A bamboo shoot 10 feet tall has a break near the top. The configuration of the main shoot and its broken portion forms a triangle. The top touches the ground 3 feet from the stem. What is the length of the stem left standing? (China, 100 BCE)
>
> A spear 20 feet long rests against a tower. Its end is moved out 12 feet. How far up the tower does the spear reach? (Italy, 1300 CE)

While the information within the sequence of problems is similar, the required mathematical results are different. If for each problem, x represents the desired unknown, the situations are quite different, as shown in figure 2.1.

Historical enrichment can be employed as reinforcement for a concept already learned. Thus, medieval Italian problems on the "rule of three" can provide practice on proportional relationships. Perhaps the simplest strategy is just to include an occasional historical problem in classroom assignments, with its date and location of origin noted. Such inclusions become an implicit reminder that mathematics has a heritage and serve as a source of self-satisfaction for students when they realize they are part of an ongoing process, the use of mathematics to solve problems.

Some historical problems easily lend themselves to "what if" scenarios. For example, several years ago in the *Mathematics Teacher*, a reader submitted his student's solution derivation for the following problem from China (100 BCE):

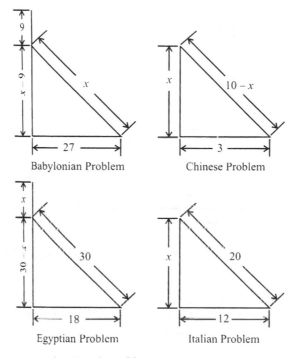

Figure 2.1. Various right triangle problems.

Given a right triangle with legs of length a and b and hypotenuse of length c, what is the length, s, of the side of the largest inscribed square using the right angle as one of its vertices? (Wikenfield 1985)

The required square is found to have side $s = ab/(a+b)$, the product of the legs over the sum of the legs (fig. 2.2).

Follow-up letters admired the problem, and an immediate "What if?" scenario was given: "What is the side of the largest inscribed square drawn along the hypotenuse?" (Lieske 1985). Another reader pointed out that the solution is half the harmonic mean between legs a and b. In still another response, I drew attention to the series of ingenious right triangle problems from which this one originated and gave the following companion problem, from the *Jiuzhang suanshu* (China, 100 BCE):

Find the radius, r, for the largest inscribed circle in this same triangle. (Swetz and Kao 1977)

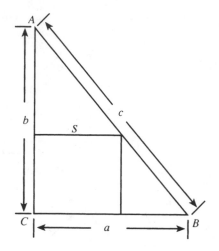

Figure 2.2. Student's solution to a 100 BCE Chinese problem published in the *Mathematics Teacher.*

In figure 2.3a the length of the hypotenuse AB is represented by c, the length of BC by a, and the length of leg CA by b. Then,

2(area $\triangle$ ABC) = ab = 4(area $\triangle$ ADO) + 4(area $\triangle$ BEO) +2(area $\square$ DOEC).
These regions can be rearranged and will be equal in area to

area $\square$ AHEC + area $\square$ DGBC + area $\square$ AHOD + area $\square$ GBEO.

Then, $ab = br + ar + cr$ and $r = ab/(a+b+c)$.

Figure 2.3b is a Chinese illustration demonstrating the geometric approach used in solving this problem.

The Chinese seemed fascinated by right-triangle relationships and devised hundreds of problems involving right triangles. For example, Li Zhi in his 1248 *Sea Mirror of Circle Measurement* presents 170 such problems, whose solutions result in a consideration of higher-degree equations. One of Li's problems is the following:

> There is a tree 135 *bu* from the southern gate of a circular city. The tree can be seen if one walks 15 *bu* from the northern gate and then 200 *bu* eastward. Find the diameter of the circular walled city.

If we allow x to be the unknown radius of the city, we find that

$$x^4 + 150x^3 + 5625x^2 - 2920320x - 788486400 = 0.$$

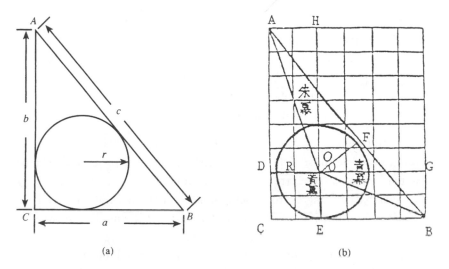

(a)   (b)

Figure 2.3. Companion problem to the one considered in figure 2.2: (a) the problem in modern terms; (b) its solution, in an illustration from a Chinese commentary on the *Jiuzhang suanshu*, chapter 9.

A correct positive root is obtained. Any class of students will ask, "How did the Chinese do it?" Compare this problem with a previous square-walled Chinese city problem, and one sees a "What if?" extension. Chinese mathematicians were still using right triangle problems well into the nineteenth century.

As we saw in chapter 1, historical problems frequently contain data that sheds light on the social and economic climate of their times. They tell the reader about the daily living conditions of the period. Consider the following pursuit problem:

> A certain slave fled from Milan to Naples going ¹⁄₁₀ of the whole journey each day. At the beginning of the third day, his master sent a slave after him and this slave went ¹⁄₇ of the whole journey each day. I do not know how far it is from Milan to Naples, but I wish to know when they overtook him. (Cardano 1539)

Here, the reader learns that slavery existed in sixteenth-century Europe. Students may then explore questions like "Who were these slaves?" and "What did slaves do in this society?"

Occasionally, problems may confront students with unfamiliar, antiquated units of measure:

If 8 *braccia* of cloth are worth 11 florens, what are 97 *braccia* worth? (Italy, thirteenth century)

I have two fields of grain. From the first field I harvest ⅔ baskets of grain per unit area. The yield of the first field exceeds the second by 500 baskets. The total area of the two fields together is 30 *sar*. What is the area of each field? (Babylonia, 2000 BCE)

"What is a *braccia*?" "What is a *sar*?" "How do these units of measure convert into modern units?" "Do they seem to be established on an understandable mathematical standard?" These are worthwhile, thought-provoking questions that can emerge from such problems, with their unfamiliar terms. If, however, unknown terms detract too much from the mathematical considerations of a problem, they can be replaced by familiar terms; for example, in the problems just considered *braccia* can be changed to "meters" and *sar* to "square rods."

The economic realities of life in the nineteenth-century United States emerge from the following problem:

A teacher agreed to teach 9 months for $562.50 and his board. At the end of the term, on account of two months absence caused by sickness, he received only $409.50.What was his board per month? (Milne 1892)

Questions that could easily follow such a problem are "What was a teacher's salary at this time?" and "How did it compare with other salaries?" The latter question could result in further research.

Consider an example of another situation, that of paying interest on a loan:

If the interest of 100 for a month be 5, say what is the interest of 16 for a year? (Bhāskara II, 1150)

Here the rate of interest is found to be 60%. In the Middle Ages, some kings, in order to finance their wars, borrowed money at the rate of 300% and more!

The human cost of warfare is illustrated in other problems:

The King of France entered into a battle and was defeated in such a way that ¼ of his soldiers were killed, ⅖ were wounded, 1000 were

32

taken prisoner and 6000 were left on the field. I want to know how many soldiers he had before he was defeated? (Italy, 1600)

After a terrible battle it is found that 70% of the soldiers have lost an eye, 75% an ear, 80% an arm and 85% a leg. What percentage of the combatants must have lost all four? (Lewis Carroll, ca. 1880)

The second problem better expresses the antiwar feeling of its Victorian author—the mathematician Charles Lutwidge Dodgson (who under the pen name Lewis Carroll also wrote *Alice in Wonderland*)—than the realities of warfare.

Of course, some problems can be enjoyed merely for their exotic whimsy:

A fox, a raccoon, and a hound pass through customs and together pay 111 coins. The hound says to the raccoon, and the raccoon says to the fox: "Since your fur is worth twice as much as mine, then the tax you pay should be twice as much!" How much should each pay? (China, 200 BCE)

The square root of half the number of bees in a swarm has flown out upon a Jessamine bush; 8/9 of the swarm has remained behind. A female bee flies about a male that is buzzing within a lotus flower into which he was allured in the night by its sweet odor but in which he is now imprisoned. Tell me, most enchanting lady, the number of bees? (India, 1150)

Still others are posed merely as riddles, tests of the reader's mathematical acuity:

A man, his wife and two sons desire to cross a river. They have a boat that will carry only 100 lbs. The man weighs 100 lbs, the wife 100 lbs and the sons each 50 lbs. How can they all cross the river using the boat? (United States, 1905)

The European origins of this problem can be traced to Alcuin of York in the year 800. As we have already seen in chapter 1, variations of this "river crossing" problem have existed in many cultures, with the boat's occupants replaced by characters at odds with one another, such as animals and their prey, cannibals and missionaries, or a protective brother transporting his virgin sisters (fig. 2.4; see Ascher 1991 for further discussion).

Figure 2.4. Illustration from fourteenth-century Italian codex *Antichissimo di Algorismo* depicting "transportation of virgins" problem.

A viable activity for students is to trace the historical migration of particular problems across cultures and centuries. One such problem is the "hundred fowls" problem, here appearing in its medieval European guise (775) and then in a twelfth-century Turkish variation.

A hundred bushels of grain are distributed among 100 persons in such a way that each man receives 3 bushels, each woman 2 bushels, and each child a half a bushel. How many men, women, and children are there?

It is known that during a day a public bath had 30 visitors who used the facility and in total paid 30 coins. If a Muslim pays half a coin for a bath, a Christian 2 coins, and a Jew 3 coins, how many Muslims, Christians, and Jews used the bath this day?

Another is the "broken tree" problem, which has appeared in many textbooks over the centuries.

A tree 100 feet high is broken in a storm and the top touched the ground 40 feet from the foot of the tree. What is the length of the portion broken off? (United States, 1905)

What is the lineage of such problems, and why are their situational and mathematical content so popular? Nice historical investigations can be structured around such questions. A problem's content often highlights applications deemed important for a particular time and place, whether it be the mercantile pursuits of Venetians, as in the first example here, or the frustrations of a British surveyor, as in the second:

A man has four creditors. To the first he owes 624 ducats; to the second, 546; to the third, 492; and to the fourth, 368. It happened that the man defaulted and escaped, and the creditors found that his goods amounted to 830 ducats in all. In what ratio should they divide this and what will be the share of each? (Tartaglia 1556)

Being employed to survey a field, which I was told was an exact geometrical square but by reason of a river running through it, I can only obtain partial measurements. I measure 9 yards from the west corner along the south side. Then sighting upon the northeast corner, I measure 18 yards along this line before turning and sighting back to the south-east corner which I find at an angle of 28° 30' from my path of previous sightings. From these measurements, determine the area of the field. (London, 1797)

Examining the contents of such problems reveals the early dependence of mathematics on real-world applications. The oldest extant problems focus on very basic human needs: the distribution of food, payment for labor, the measurement of land, and the payment of taxes. Later, as social movements develop and scientific and technological innovations take place, those movements and innovations also become subjects in mathematical problems.

In contrast to problems that were published as instructional aids, there are problems that were conceived and issued as direct challenges to the mathematical ability and ingenuity of the readers. At times, authors have directed these challenges at selected colleagues. Modern students can obtain an added degree of satisfaction by successfully solving problems that baffled their ancestors, even ancestors that were noted mathematicians in their time. Here are a few such challenge problems.

From Christiaan Huygens to Gottfried Leibniz, 1672:

What is the sum of the reciprocals of the triangular numbers?

From Pierre de Fermat to John Wallis, about 1657:

> Draw a rectangle ABCD with sides b and $a = b/\sqrt{2}$, and raise a semi-circle on one of the sides $AB = b$. Now pick any point E on the semi-circle. Join E to the corners C and D of the rectangle. Call U and V, respectively, the points of intersection of ED and EC with AB. Prove that $(AV)^2 + (BU)^2 = b^2$.

From the pages of the London journal *Ladies Diary* to its readership in 1755:

> Two circles of radius 25 feet intersect so that the distance between their centers is 30 feet. What is the length of a side of the square inscribable within the space defined by the intersecting arcs?

In such instances, an ensuing discussion should focus on the issue of what different mathematical tools and procedures we now have that help in obtaining the desired solutions, tools and procedures our ancestors did not possess. A most fruitful challenge problem from the Middle Ages is

Figure 2.5. The "pons asinorum" problem as given in Barrow's 1665 edition of *Euclid's "Elements."*

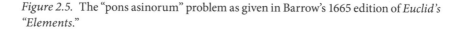

PROP. V.

Iſoſcelium triangulorum ABC *qui ad baſim ſunt anguli* ABC, ACB *inter ſe ſunt aquales. Et productis aqualibus rectis lineis* AB, AC *qui ſub baſe ſunt anguli* CBD, BCE *inter ſe aquales erunt.*

Accipe AF = AD, & junge CD, ac BF.

Quoniam in triangulis ACD, ABF, ſunt AB = AC; & AF = AD, angulusq; A communis, erit ang. ABF = ACD; & ang. AFB = ADC, & baſ. BF = DC; item FC = DB. ergò in triangulis BFC, BDC erit ang. FCB, = DBC. Q.E.D. Item ideo ang. FBC = DCB. atqui ang. ABF = ACD. ergò ang. ABC = ACB. Q.E.D.

a 3 I.
b I. p ſt.
c byp.
d conſtr.
e 4. I.
f 3 ax.
g 4. I.
h pr.
k 3. ax.

Corollarium.

Hinc, O nne triangulum æquilaterum eſt queq; æquiangulum.

36

the *pons asinorum*, or "bridge of asses," exercise. The problem is to provide a proof for the theorem that the base angles of an isosceles triangle are equal. The approved proof of the time is shown in figure 2.5, reproduced from Isaac Barrow's 1665 edition of *Euclid's "Elements."* It is written in Latin, the language of scholarly mathematics at this time; however, a student with some knowledge of geometry should be able to decipher and follow it—a problem of mathematics understanding in itself.

Why do you think the proof was done this way? Give a "better" proof for this theorem. In this instance, just what does "better" mean? Why is it called "the bridge of asses"? These are just a few of the questions that can be constructed around this problem. The reader should try to extend this list.

Conclusion

The use of historical problems is an excellent way of enriching instruction in mathematics. These problems provide a context of good problem-solving and also motivate students to learn, both because they help students realize that math has had practical uses for millennia and because they lead students to pose additional questions about the past.

One question remains: Where can teachers obtain such problems? They can be obtained from old textbooks, survey histories and specialized histories of mathematics, and specific online collections such as that on the MathDL website (*Loci*: Covergence section), sponsored by the Mathematical Association of America. The bibliography at the end of this book can supply many more problems.

Since we began with a problem, let us end with a problem, one from fifteenth-century Italy that was solved without a use of trigonometric functions:

A circle of radius 4 units is inscribed in a triangle. A point of tangency of the circle with the side of the triangle divides the side into lengths of 6 and 8 units. What are the lengths of the remaining two sides of the triangle?

How did they solve it?

3 Ancient Babylonia (2002–1000 BCE)

Cuneiform tablet YBC 7289, one of the most historically significant mathematical clay tablets ever found. Dated to the Old Babylonian Period (ca. 1800–1600 BCE), it was acquired by the American millionaire J. P. Morgan. He contributed it to Yale University in 1912.

It is believed that the grain that we know as wheat first appeared 10,000 years ago in the Middle East. This food was cultivated on the fertile floodplains of the Tigris and Euphrates Rivers. Here in Mesopotamia, the "land between the rivers," an agricultural society prospered and grew into what would be known as the Babylonian civilization. The Babylonians soon became merchants and traders and extended their influence over a large geographical region. The needs of this empire warranted the development of sciences. Babylonians developed very accurate systems of mathematics and astronomy. Their number system was based on counting by 60s; our use of 60 minutes in an hour and 360° in a circle evolved from Babylonian tradition. Babylonian astronomy, serving the needs of an agricultural society, was very exact. As a result, other peoples followed Babylonian astronomical procedures, including performing calculations in base 60.

Using a system of writing known as cuneiform, ancient Babylonian scribes recorded their results on clay tablets. The clay table in the photograph above contains a student exercise in which the student is asked to find the length of the diagonal of a square. Thousands of such clay tablets from ancient Babylonia have been discovered, most of them at the sites of the ancient cities of Nippur and Susa. Major collections of these tablets are held by the University of Pennsylvania, Yale University, Columbia University, and the Louvre in Paris. It is estimated that about 400 of these tablets concern mathematics, but few have been translated. A modern researcher, Eleanor Robson of Oxford University, has undertaken the challenge of exploring them and learning more about ancient Babylonian mathematics. Her most recent book on this subject is *Mathematics in Ancient Iraq* (Robson 2008).

For computational ease in the following problems, numbers originally given in sexagesimal form (base 60) have been converted to base 10. To preserve historical interest, some Babylonians units of measure are retained, usually in a self-explanatory manner. It should be remembered that the Babylonians had extensive and complex systems of metrology. As an added facet of historical and mathematical interest, Babylonian algebraic and geometric terminology is presented in several problems, such as problems in which an unknown was perceived as the length of a line, a product of two unknowns, the area of a rectangle, or when appropriate, a square. Contemporary student problem solvers will sometimes have to translate the Babylonian terminology to present-day algebraic form and proceed accordingly.

For readers used to doing their computations in base 10 notation, the concept of a sexagesimal arithmetic may seem very strange. But mathematically speaking, 60 is a nice number. It has many integral divisors and is useful in the tasks of dividing up commodities. Some Babylonian experts believe that sexagesimal computation was first used in the weighing and measurement of grain and was eventually adopted into mathematical practices in general. Let us briefly examine how the system works. In sexagesimal numbers every place value can be occupied by up to two digits representing a value less than 60. This place value holder is the coefficient for a power of 60. The popular convention for writing translated cuneiform sexagesimal numerals from Babylonia is to use commas to mark sexagesimal places. For example, consider the translation of a number written as

$$7, 23, 41 \text{ (base 60).}$$

We would interpret this in base 10 as 41×60^0 (units) 23×60^1 and 7×60^2. This would give

$$41 + 1380 + 25{,}200 = 26{,}621 \text{ (base 10).}$$

Converting from base 10 to base 60, one reverses the operations, thus:

$$4389 \text{ (base 10)} = 1, 13, 9 \text{ (base 60).}$$

A historically unique feature of the Babylonian numeration system is that it accommodated fractions. Using the conversion/translational system, a sexagesimal point is represented by a semicolon. Thus, 25; 12, 17 represents $(25 \times 60^0) + (12 \times 60^{-1}) + (17 \times 60^{-2})$, or 25.2047. In going from a decimal number to a sexagesimal equivalent, the process is again reversed, so that 15.812 (base 10) = 15; 48, 44 (base 60). Try converting some common fractions ($\frac{1}{4}$, $\frac{5}{8}$, $\frac{11}{16}$, $\frac{17}{125}$) to their sexagesimal forms.

Problems

1. A wooden beam is stood vertically against a wall. The length of the beam is 30 units. If the top of the beam slides down the wall 6 units, how much does the lower end slide out horizontally along the ground?

2. In a circle whose circumference is 60 units, a chord is drawn, forming a segment whose *sagitta* is 2 units. What is the length of the chord?

3. In a circle whose circumference is 60 units, a chord 14 units long is drawn constructing a segment of the circle. What is the length of the sagitta of this segment?

4. The sum of the area of two squares is 1525. The side of the second square is two-thirds that of the first plus 5 units. Find the sides of each square.

5. A circle contains an inscribed triangle whose sides are 50, 50, and 60 units, respectively. What is the radius of the circle?

6. Let the width of a rectangle measure one-fourth less than its length. Let 40 units be the length of the diagonal. Find the length and width of this rectangle.

7. Given an isosceles trapezoid with bases 17 *cubits* and 7 cubits and a height of 12 cubits, find the length of the transversal, drawn parallel to the bases that divides the trapezoid into two equal areas.

8. There are two silver rings; $1/7$ of the first and $1/11$ of the second ring is broken off, so that what is broken off weighs one *shekel*. The first that is diminished by $1/7$ weighs as much as the second diminished by its $1/11$. What was the weight of the silver rings originally?

9. I have two fields of grain. From the first field, I harvest two-thirds of a bushel of grain per unit area; from the second, one-half a bushel per unit area. The yield of the first exceeds the second by 50 bushels. The total area of the two fields together is 300 units. What is the area of each field?

10. A little rectangular canal is to be excavated for a length of 5 km. Its width is 2 m, and its depth is 1 m. Each laborer is assigned to remove 4 m³ of earth, for which he will be paid one-third of a basket of barley. How many laborers are required for the job, and what are the total wages to be paid?

11. It is known that the digging of a canal becomes more difficult the deeper one goes. In order to compensate for this fact, differential work allotments are computed: a laborer working at the top level is expected to remove $1/3$ *sar* of earth in one day, while a laborer at the middle level removes $1/6$ sar, and one at the bottom level, $1/9$ sar. If a fixed amount of the earth is to be removed from the canal in one day, how much digging time should be spent at each level?

12. A granary of barley contains 2400 *gur*, where 1 gur equals 480 *sila*. If workers are to receive 7 sila of grain for a day's work, how many men can be paid from this granary?

13. I have added up 7 times the side of my square and 11 times its area and obtained 6 ¼. What is the side of my square?

14. Two-thirds of two-thirds of a certain quantity of barley is taken, and we obtain 100 units. What was the original quantity?

15. Ten brothers receive an inheritance of 100 shekels to be divided among them. The money is to be divided such that each brother receives an excess equal to that given his previous brother, so that each brother gets more than his previous brother by an equal amount. It is known that the eighth brother receives 6 shekels. What is the differential payment between brothers?

16. A wall has length of 60 units. The top is ½ unit, the base is 1 unit, and the height is 6 units. What is the volume of the wall?

17. A woman weaves a textile that is to be 48 rods long. In one day, she weaves ⅓ rod. In how many days will she cut the textile from the loom [be finished]?

18. A circle has a circumference of 60 rods. I descend two rods into the circle. What is the length of the dividing line I reached (the chord)? [Note: For this calculation let $\pi=3$.]

19. A square plus its side equals ⁴⁵⁄₆₀. What is the side of the square?

20. A square minus its side equals 870 units. What is the side of the square?

21. The sum of two squares equals 1525 units. The side of one square equals two-thirds the side of the other square plus 5 units. What are the sides of my squares?

22. I have a square and added four times the length of its side and obtained ²⁵⁄₃₆. What is the side of my square?

23. The sum of my two squares is 1300. The side of one exceeds the side of the other by 10 units. What are the sides of my squares?

24. The sum of two squares is 1300. The product of their two sides is 600. What are the sides?

25. Given a rectangular trench, whose surface area 7½ sar. Its volume is 45 sar. I summed the length and width and obtained 6½ rods. What is the length, width, and depth of this trench? [1 sar area=1 square rod; 1 sar volume=1 sar area times 1 cubit.]

26. A canal is 5 rods long, 1½ rods wide, and ½ rod deep. Workers are assigned to dig 10 *gin* of earth for which task they are paid 6 sila of grain. What is the area of the surface of this canal and its volume? What are the number of workers required and their wages? [1 rod=12 cubits; 1 cubit=60 gin.]

27. A siege ramp is to be built to attack a walled city. The volume of earth allowed is 5400 sar. The ramp will have a width of 6 rods, a base length of 40

rods, and a height of 45 *kus*. Construction of the ramp is incomplete; an 8-rod gap is left between the end of the ramp and the city wall. The height of the uncompleted ramp is 36 kus. How much more earth is needed to complete this ramp? [1 kus=1 cubit; 1 rod=12 cubits.]

28. Try a problem, as translated directly from a cuneiform tablet. Remember that numbers are given in sexagesimal form.

Given a crescent moon, the arc of its circle is 1 00, the dividing line 50. What is the area? You: By what does 1 sixty, the circle, exceed 50? The excess is 10. Multiply 50 by 10, the excess. You obtain 8 20. Square 10, the dividing line. You obtain 1 40. Take away 1 40 from 8 20. The result is 6 40. The area is 6 40. This procedure is leading you through an approximation for the area of a circular segment. Do you see how this approximation works? Is it a good approximation? If not, how could you make it a better approximation?

The next two problems were given as scribal exercises to students during the Old Babylonian era. They are approximately 4000 years old. See how well you can do with these ancient problems.

29. The side square is 1 *cable*. Inside it, I drew 16 wedges. What is the area of a wedge? [1 cable=10 rods; 1 rod=12 cubits.]

30. The side square is 1 cable. Inside it are 4 wedges, 16 barges, and 5 cows' noses. What is the area of a cow's nose?

WHAT ARE THEY DOING?

Using Iteration to Approximate the Square Root of 2

The illustration that begins this chapter is a photo of a cuneiform tablet, YBC 7289, in the Yale University Babylonian Collection. It is one of the most dramatic and revealing mathematical artifacts ever found. On close examination it appears to be a geometric sketch of a square with two intersecting diagonals accompanied by cuneiform writing as shown in figure 3.1a. When the cuneiform markings are translated, they give the numbers shown in figure 3.1b.

If we examine the numbers written above the horizontal diagonal and place a sexagesimal point between the 1 and 24, we can interpret this number sequence in the following way.

$$1 + \frac{24}{60} + \frac{51}{60^2} + \frac{10}{60^3} = 1 + \frac{2}{5} + \frac{51}{3600} + \frac{1}{21,600}$$
$$= 1 + 0.4000000 + 0.0141666 + 0.0000462$$
$$= 1.4142129$$

These same numbers multiplied by the length of the side of the square (30 units) gives a sexagesimal value of 42; 25, 35. These are the numbers

Figure 3.1 (a) Archeological sketch of YBC 7289 by A. Aaboe published in *Episodes from the Early History of Mathematics*. (b) Translation of the inscription.

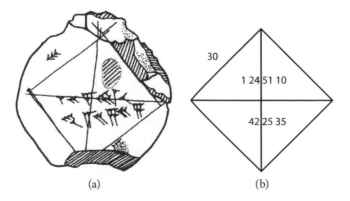

(a)   (b)

appearing under the diagonal and would seem to represent the length of these diagonals. The number 1.4142129 is an approximation to $\sqrt{2}$. My calculator value for $\sqrt{2} = 1.41421356$. The ancient Babylonians therefore achieved an accurate estimate for $\sqrt{2}$ within 0.02% error! How did they do it?

While no one knows exactly how they did this, we suspect it was done by "the divide and average" method. This method works the following way:

1. You are asked to approximate $\sqrt{x}$, for which you make a guess, g_1.
2. You then find $x/g_1 = e_1$.
3. Then let $g_2 = (g_1 + e_1)/2. \ldots$
4. Now return to the first step and repeat the process using g_2. The process is repeated until sufficient accuracy is achieved.

As an example, let us estimate $\sqrt{139}$.

1. I make a first guess of 11.
2. Then, $139/11 = 12.63636$.
3. $(11 + 12.63636)/2 = 11.81818$.
4. $139/11.81818 = 11.76154$.
5. $(1.76154 + 11.81818)/2 = 11.7898$.

Thus after two guesses, or iterations, I obtain a desired answer:

$$(11.7898)^2 = 138.999.$$

The process of mathematical iteration—that is, of obtaining a series of increasingly accurate approximations to a desired mathematical solution—is an important aspect of modern computing. Here we see that this process was used to solve problems 3000 years ago.

4   Ancient Egypt

Wall painting from the tomb of Menna, built in the period 1420–1411 BCE in the Valley of the Kings. Menna was a chief scribe in the eighteenth dynasty who supervised the measuring of fields and the collection of taxes. He bore the honorary title "Scribe of the Fields." This tableau shows agricultural scenes of this period and reflects on the uses of mathematics. The upper level depicts the measuring of the grain fields; on the middle level we see the measuring of grain harvested and the computation of taxes; the lower level shows the paying of taxes which are then carried away on the right side.

People settled along the banks of the Nile River more than 5000 years ago. They used the river for irrigation and developed a thriving agricultural society. The region in which they settled became known as Egypt and the people, the Egyptians. While some information concerning ancient Egyptian mathematics can be gleaned from inscriptions on stone tombs and monuments, most of our information on the subject is found on less durable material: papyrus and leather scrolls. Papyrus was a form of paper made from reeds that grow in abundance along the banks of the Nile River. Using ink, Egyptian scribes wrote information on papyrus scrolls in hieroglyphics, either in pictograms or a hieratic script. Both the ravages of time and human conquests have left few Egyptian documents from which a mathematical testimony can be found. The principal sources of information on mathematics are the Moscow Mathematical Papyrus (1850 BCE), the Rhind Mathematical Papyrus (1650 BCE), the Rollin Mathematical Papyrus (1350 BCE), and the Harris Mathematical Papyrus (1167 BCE). All these papyri represent collections of problems that dealt with the running of the Egyptian kingdom.

Problems

1. [Given] a quantity: its two-thirds, one-half, and one-seventh are added together, giving 33. What is the quantity?

2. A riddle problem from the Rhind papyrus:

There is an estate that contains 7 houses; each house has 7 cats; each cat catches 7 mice; each mouse eats 7 *spelt* of seeds; each spelt was capable of producing 7 *hekats* of grain.

How many things were in the estate? The answer is given as 19,607; is this correct? Show that this is the correct answer for the sum of the geometric series $7 + 49 + 343 + 2401 + 16,807$.

3. Divide 100 loaves of bread among 10 men, including a boatman, a foreman, and a doorkeeper, who all received double proportions. What is the share of each?

4. How many cattle are in a herd with two-thirds of one-third of them equal to 70, the number given as tribute to the owner?

5. Suppose a scribe tells you that four overseers have drawn 100 great quadruple hekats of grain, and their work gangs consist of 12, 8, 6, and 4 men. How much grain does each overseer receive?

6. Given: a circle with an inscribed equilateral triangle. The triangle has an area of 12 square units. What is the area of the circle?

7. Given a circular conduit 100 *cubits* in length, 3 cubits in diameter at its base, and 1 cubit in diameter at its top, determine its volume.

8. A rectangular plot of land is 60 cubits square; the diagonal is 13 cubits. How many cubits does it take to make the sides?

9. An erect pole, 10 cubits in length; as its base moved outwards, 6 cubits determined the new height and the distance the top of the pole has been lowered.

10. If it is said to you, "Have sailcloth made for the ships," and it is further said, "Allow 1000 cloth cubits for one sail and have the ratio of the height of the sail to its width be 1 to 1 ½," what is the height of the sail?

11. A quantity plus 17 of it becomes 19. What is the quantity?

12. A quantity together with its two-thirds has one-third of its sum taken away to yield 10. What is the quantity?

13. The sum of a certain quantity, together with its two-thirds, its half, and its one-seventh, becomes 37. What is the quantity?

14. You are told that the area of a square of 100 square cubits is equal to the area of two smaller squares; the side of one is one-half plus one-fourth of the other. What are the sides of the squares?

15. Divide 10 hekats of barley among 10 men so that the common difference is ⅛ hekat.

16. Divide 100 hekats of grain among five men so that the common difference is the same and so that the sum of the two smallest shares is one-seventh the sum of the three largest.

17. Find the volume of a cylindrical granary of diameter 10 cubits and height 10 cubits.

18. Given a circular area whose diameter is 9 hekats. What is its area?
The Egyptian directions for solving this problem are as follows:

You shall subtract from the diameter ⅑ of its length; the remainder is 8. You shall multiply 8 by 8; the result is 64. This is the amount of the area.

How are the Egyptians approximating the area of a circle? If they were using our modern formula for this computation, what would be their value for π?

19. Given four numbers, their sum is 9900. Let the second exceed the first by 17, let the third exceed the sum of the first two by 300, and let the fourth exceed the sum of the first three by 300. Find these numbers.

20. Given a pyramid 300 cubits high, with a square base 500 cubits on a side, determine the distance from the center of any side to the apex.

21. Find the height of a square pyramid with a *seqt* of 5 hands and 1 finger per cubit and a base of 140 cubits on one side.

[Note: The Egyptian measure *seqt* is the slope of the side of a pyramid. It is formed by determining the ratio of the run divided by the rise, that is, by giving the horizontal distance of the oblique face from the vertical for each unit of height. The vertical unit was taken as the cubit and the horizontal unit as a hand, where 7 hands = 1 cubit and 4 fingers = 1 hand.]

22. A measure of cloth that is 7 cubits in height and 5 cubits in width amounts to 35 cloth cubits. Take off 1 cubit from the height and add it to the width so that the area remains the same. What measurement is added to its width?

23. A pyramid has a base of 360 cubits and a height of 250 cubits. What is its seqt?

24. A cobbler can cut leather for ten pairs of shoes in one day. He can finish five pairs of shoes in one day. How many pairs of shoes can he both cut and finish in one day?

25. In ancient Egypt the quality of products made with grain, such as bread and beer, was measured by a unit called a *pesu*, where the value of a pesu is computed as follows:

$$\text{Pesu} = \frac{\text{Quantity of bread or beer}}{1 \text{ hekat}^3 \text{ of grain}} \qquad 4.1$$

How much beer can be made with a strength of 2 pesu from 10 hekats[3] of grain?

WHAT ARE THEY DOING?

Computing the Volume for the Frustum of a Pyramid

When archaeologists translate ancient Egyptian papyri, they often take the content—usually written in hieratic, a cursive script that was used by scribes—and first translate it into hieroglyphics, a pictograph form, then finally into a modern language. Figure 4.1 is a facsimile of problem 14r

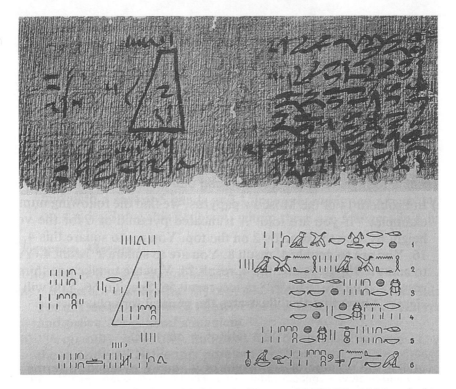

Figure 4.1. Fragment of Moscow Mathematical Papyrus (1850 BCE) with hieroglyphic translation at the bottom.

taken from the Moscow Mathematical Papyrus (1850 BCE). At the bottom of the illustration is a translation of the text into hieroglyphics.

In English this problem reads as follows:

Method of calculating a square pyramid.
If you are told of a square pyramid of 6 as height, of 4 as lower side, and 2 as upper side.
You shall square these 4. 16 shall result.
You shall double 4. 8 shall result.
You shall square these 2. 4 shall result.
You shall add the 16 and the 8 and the 4. 28 shall result.
You shall calculate ⅓ of 6. 2 shall result.
You shall calculate 28 times 2. 56 shall result.

Look, belonging to it is 56.

What has been found by you is correct.

To help us understand the directions, let us use some general notation. Visualize a frustum of a square pyramid, with height $= h$, the length of the upper base $= a$, and the length of the lower base $= b$. Then the first line of the directions tells us that $h=6$, $b=4$, and $a=2$. Proceeding to the next steps, we have the following:

$$4^2 = 16, \ b^2 = 16$$
$$2 \times 4 = 8, \ ab = 8$$
$$2^2 = 4, \ a^2 = 4$$
$$16 + 8 + 4 = 28$$
$$\frac{6}{3} - 2, \ \frac{h}{3} = 2$$
$$2 \times 28 = 56 \Rightarrow \text{Volume} = \frac{h}{3}(b^2 + ab + a^2)$$

The volume of the frustum of the pyramid is 56 [cubic units].

The scribes are using the correct modern formula for the volume of the frustum of a square pyramid. How did they obtain this formula? We believe they derived it by a method of dissection. As an exercise, make a sketch of the frustum of a square pyramid and label its dimensions as above: a, b, and h. Using these dimensions, break up or dissect the frustum into several parts (as shown in fig. 4.2): a central rectangular prism, two rectangular prisms formed by the side pieces in pairs, and finally a pyramid formed by the four corner pieces. Using the dimensions you assigned, compute the volume of each one of these pieces. (You should know the formulae required for these three volumes.) Algebraically add the expressions you obtained and simplify the result to see if you arrive at the formula given above.

Figure 4.2. Dissection of a frustum into simpler geometric solids.

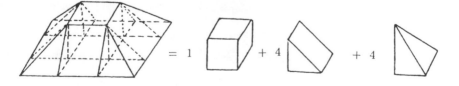

Throughout the history of mathematics, problems have been solved by this technique: A problem whose solution is required is broken down into a series of simpler problems whose solutions are known. In a sense, the solution for the large, difficult problem becomes the sum of the solutions for small, easier problems. In this manner, a problem solver is exchanging a difficult problem for a series of simpler problems that he or she can solve.

5 Ancient Greece

Pythagoras contemplates the relationship of numbers to the universe.

History recognizes Thales of Miletus (ca. 624–546 BCE) as the first Greek mathematician. He traveled widely in the Mediterranean region and is believed to have been a merchant. In Egypt, he observed priest-surveyors using geometric principles to establish land boundaries. In Babylonia, he acquired the techniques of observing the heavens—astronomy. Thales brought this knowledge back to Greece and became a teacher of the mathematics he had learned, and developed new mathematical ideas. Admired as a philosophical and intellectual science, mathematics evolved along more formal and abstract ways in Greece than in other areas of the world at this time. The Greeks viewed mathematics as divided into two parts: *arithmetica*, the abstract study of numbers, their properties and relationships; and *logistica*, applied mathematics, the solving of simple everyday problems. Greek scholars studied *arithmetica*, while *logistica* was left for slaves, merchants, and craftsmen to undertake. The following Greek problems reflect this intellectual bias.

Problems

1. Find two numbers with sum 20 and when squared their sum is 208.

2. Given, four integers where, if added together three at a time, their sums are 20, 22, 24, and 27. What are the integers?

3. Heron of Alexandria (ca. 75) wrote on many aspects of applied mathematics. In his work *On Measurement,* he presents a formula for the area of a triangle given the lengths of its three sides, a, b, and c.

$$\text{area} = [s(s-a)\,(s-b)\,(s-c)]^{1/2},$$
$$\text{where } s = \frac{a+b+c}{2}.$$

Prove that Heron's formula is correct.

4. In the *Book of Lemmas* (ca. 250), Archimedes introduces a figure that, due to its shape, has historically been known as the "shoemaker's knife," or *arbelos.* If in a given semicircle with radius R and diameter AB, two semicircles with radii r_1 and r_2, where $r_1 \neq r_2$ and $r_1 + r_2 = R$, are constructed on diameter AB so that they meet at point C on AB, then the region bounded by the three circumferences is called an arbelos.

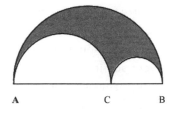

A   C   B

The arbelos fascinated Archimedes with its mathematical proprieties. Let us explore some of these properties:

(a) Prove that the length of arc *AC* plus the length of arc *CB* equals the length of arc *AB*.

(b) Prove: If a perpendicular line is constructed from *C* intersecting the arc *AB* at point *P*, then *PC* is the diameter of a circle whose area equals that of the arbelos.

(c) Complete the lower half of the circle with diameter *AB*. Let the midpoint of the arc of this lower semicircle be *Q*, the midpoint of arc *AC* be *M*, and the midpoint of arc *CB* be *N*. Prove that the area of quadrilateral *MCNQ* is equal to $r_1^2 + r_2^2$.

For further information on the arbelos and its mathematical properties, go online and research this topic.

5. Find two numbers such that their difference and the difference of their cubes are equal to two given numbers.

6. Four waterspouts are filling a tank. Of the four spouts, one can fill the tank in one day, the second takes two days, the third takes three days, and the fourth takes four days. How long will it take all four spouts working together to fill the tank?

7. Two friends were walking. One said to the other, "Give me 10 coins, and I will have three times as much money as you." And the other said, "If you give me the same amount, I will have five times as much as you." How many coins does each have?

8. Find two numbers with the sum 20 and when they are squared, this sum becomes 208.

9. Given four integers, where, if added together three at a time, their sums are 20, 22, 24, and 27. What are the integers?

10. Demochares has lived a fourth of his life as a boy, a fifth as a youth, and a third as a man, and has spent 13 years in his old age. How old is he?

11. Bricklayer, I am in a hurry to finish this house. Today is cloudless, and I do not require many more bricks, for I have all that I want except 300. You alone in one day could make this amount, but when your son quit working, he had finished 200, and your son-in-law had quit when he had made 250. Working all altogether, in how many days can you make these bricks?

12. Two people are having a conversation. The first says, "Give me 2 *minae*. I will have twice as much as you." The second responds, "And if I receive the same amount from you, I will have four times as much as you." How many minae does each person have?

13. Three people are having a conversation. The first person says, "I am equal to the second and one-third of the third." The second person says, "I am equal to the third and one-third of the first." The last person exclaims, "And I, one-third of the second plus 10." What is the value of each person?

14. An old man speaks to a child. The man asks, "Where have all the apples gone, my child?" The child responds, "Ino has taken one-third, and Semale one-eighth, and Antone went off with one-fourth, while Agrave snatched from me and carried away one-fifth. Thirty apples are left. I swear, by the god Cypris, I have only this one." How many apples were there originally?

15. The inscription on the tomb reads: "This tomb holds Diophantus. Ah, how great a marvel!" The inscription then tells the length of his life as follows: God granted him to be a boy one-sixth of his life, and adding one-twelfth part of this, he clothed his cheeks with down. He lit the light of wedlock after one-seventh part of his life, and after 5 years in his marriage he granted him a son. Alas, late-born child; after reaching one-half the measure of his father's life, cruel fate took him. After consoling his grief by the science of mathematics for 4 years, Diophantus ended his life. Determine the number of years for each respective event in the life of Diophantus.

16. The three Graces were carrying baskets of apples, and each basket contained the same number. The nine Muses met them and asked each for apples, and they gave the same number to each Muse, and the nine and the three each had the same number. Tell me how many apples they gave away, and how they all had the same number.

[Note: This problem results in an indeterminate equation. Choose the smallest answer.]

17. A riddle: I am a brazen lion. My spouts are my two eyes, my mouth, and the sole of my right foot. My right eye fills a jar in 2 days, my left eye in 3

56

days, and my foot in 4. My mouth is capable of filling it in 6 hours. Tell me how long all four spouts together will take to fill the jar. [Assume 1 day = 12 hours.]

The following lament is a paraphrase of a problem given in the Greek Anthology. This collection of 46 problems was assembled by the grammarian Metrodorus (ca. 500 CE).

18. After praying for a just increase in my fortunes of gold, I have nothing. I gave 40 *talents* of gold under evil auspices to my friends in vain, and I see my enemies in possession of a half, a third, and an eighth of my fortune.

How many *talents* did this man once have?

19. Find two square numbers whose difference is a given number—for example, 60. [Hint: Consider x^2 and $(x + k)^2$].

20. Given a circle with equal chords, prove that the equal chords determine equal angles with a circumference of the circle.

21. Prove that the angle inscribed in a semicircle is a right angle.

22. If two straight lines are parallel and a straight line intersects them, prove that the alternate interior angles formed are equal.

23. Archimedes was fascinated by the relationship of the volume of three solids: a right circular cylinder with radius R and height R; a hemisphere with radius R; and a circular cone with radius R and height R. What is the relationship of the volumes of these three solids?

24. Prove that the external angles of a polygon will equal four right angles.

25. Prove that the base angles of an isosceles triangle are equal.

26. When a surveyor measures a triangular piece of land, he usually measures the length of the three sides of the triangle. From this data, at times, you must obtain the area of the triangle. Heron of Alexandria (ca. 75) developed a formula for this situation: Given a triangle with sides of length a, b, and c and whose semi-perimeter is given by s, then the area of the triangle, A, is given by the expression

$$A = \sqrt{s(s - a)(s - b)(s - c)}.$$

Use Heron's formula to find the area of a triangle with sides 7, 6, and 3.

Three classical problems of Greek antiquity are the following:

- (a) the tri-section of the angle;
- (b) quadrature, or squaring of the circle—that is, finding a circle with the same area as a given square;
- (c) the construction of a cube with double the volume of a given cube.

These three problems were to be done by construction, using only a straight edge and compass. Over the centuries, many attempts have been made to solve these problems. The following three problems reflect such attempts, particularly the squaring of the circle.

27. In the ancient Indian mathematical works called the Sulbasutras [cord-stretching manuals], a method is given for constructing a circle with the same area as a given square. Perform this construction of "circling the square"; then calculate the area of the two figures and compare your results.

(a) Draw a square.

(b) Circumscribe the square with a circle.

(c) Construct a perpendicular bisector from the center of the square O through its upper side. The bisector intersects the upper side at point P and the circumscribing circle at point M.

(d) Divide $\overline{PM}$ into thirds. Let the division points be K and N, where K lies closer to point P.

(e) Use $\overline{OK}$ as a radius to construct a circle. This circle is supposed to have the same area as the original square.

Does it?

28. Euclid, in book 2, proposition 14, of his *Elements*, gives the following construction for squaring a rectangle: Given the rectangle *ABCD*, side *AB* is extended the length *CB* to point *E*. *AE* is used as a diameter to construct a semicircle. Side *BC* is now extended to meet this semicircle at point *F*. $\overline{BF}$ now supplies the side for the required square *BFGH*.

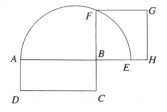

On a blank sheet of paper, construct a rectangle and carry out the above construction to find a square of the same area as the rectangle. Use your knowledge of algebra and geometry to prove that the square and the rectangle have the same area.

29. The Greek mathematician Hippocrates of Chios (ca. 440 BCE), in his attempts to square the circle, invented a theory of *lunes*. The word *luna* in Latin means "the moon." Lunes are geometric figures formed by the intersection of two circular arcs. These figures or regions look like pieces of the moon. Hippocrates discovered that one could find a particular lune with the same area as a semisquare, and that this relationship could be obtained by geometric constructions. Carry out Hippocrates' construction as follows and prove that the lune and the semisquare have the same area.

Construct a line segment AB. Using $\overline{AB}$ as a diameter, construct a semicircle. Mark the midpoint of $\overline{AB}$, C, and draw $\overline{AC}$ and $\overline{CB}$. The isosceles triangle ACB forms a semisquare. Complete the square and label the fourth vertex, D. Using D as the center of a circle, draw an arc connecting A and B. The region ACB forms a lune. Now show that the area of this lune equals the area of triangle ACB.

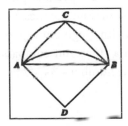

30. In their exploration of number relationships, the Greeks employed several procedures. One of these procedures was to consider the mean relationship between any two given numbers, a and b. For this task, they developed three principal means: the arithmetic mean, M_A; the geometric mean, M_G; and the harmonic mean, M_H.

These means are defined as follows:

$$M_A = \frac{(a+b)}{2},$$

$$M_G = \sqrt{ab},$$

$$M_H = \frac{2ab}{(a+b)}.$$

Order these three means according to size.

31. Find three numbers such that a product of any two added to the square of the third results in a square.

WHAT ARE THEY DOING?

Constructing a Sequence of Irrational Lengths

The ancient Pythagoreans sought harmony and consistency in numbers. Numbers represented permanency; they could he be depended upon. The motto for this mystical brotherhood was "All is number." Everything depended on whole numbers. All quantities could be compared by whole numbers or ratios of these numbers. But then a number was discovered that was incommensurable—that is, could not be compared as a ratio of two numbers—and the discovery caused a "logical scandal" among the Pythagoreans. This scandal was especially profound in that this number appeared in a very simple situation: the measurement of the diagonal of a square with unit sides. The line segment representing this number could be easily constructed with a ruler and compass, and yet, theoretically, it could not be measured. The number was $\sqrt{2}$.

Figure 5.1. Theodorus's demonstration of the irrationality of the non-square integers from 3 to 17.

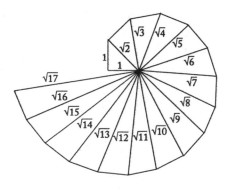

This knowledge of the existence of irrational numbers challenged sacred mathematical beliefs. But perhaps $\sqrt{2}$ was an anomaly; perhaps there was only one such number. Alas, this was not to be the case. Theodorus of Cyrene (ca. 425 BCE) demonstrated that there were many such numbers. He constructed some of them with a ruler and compass as line segments using the Pythagorean principle and stopped at the square root of 17, but his technique could be continued indefinitely.

Study the diagram in figure 5.1, and when you understand how it is devised, construct some more line segments representing irrational numbers.

Since the time of the Greeks, mathematicians have studied irrational numbers in depth. They have discovered that the irrational numbers are dense within the rational numbers. between every two rational numbers an irrational number exists. There are an infinite number of them. Also, in seeming contradiction, there are more irrational numbers than rational numbers. Important numbers such as π and e have been found to be irrational.

6 Ancient China

Frontispiece from *Suanfa tongzong* (General source of computational methods) (1592). Entitled "Discussions on difficult problems between master and pupil," it is the only known depiction of a Chinese computing board.

Like Mesopotamia and ancient Egypt, ancient China was also a "hydraulic society." It developed in fertile river valleys that allowed for agriculture. But the rivers, mainly the Yangtze and the Yellow River, were subject to flooding; and water-control dikes and irrigation systems were necessary for survival of the human settlements. Responsibility for the construction and maintenance of these systems fell to the government and its bureaucracy. Eventually, this government consisted of an emperor and imperial departments manned by court scholars. The two most important scientific disciplines employed to maintain the empire were mathematics, needed for construction and tax collection, and astronomy, to predict the agricultural growing cycles.

The few mathematical handbooks that survive contain problems that demonstrate the applied nature of early Chinese mathematics. The most important of these books is *The Nine Chapters on the Mathematical Art* (ca. 100 BCE), which served Chinese mathematical needs for hundreds of years and was widely adopted in neighboring countries such as Japan and Korea.

The problems that follow contain several traditional Chinese measures. Where necessary, some conversion relations have been given. Still, it would be a worthwhile student exercise to work out the relationships among them and see how they compare with our modern measures.

Problems

1. A wooden log is encased in a wall. If we cut part of the wall away to a depth of 1 inch, the width of the exposed log measures 10 inches. What is the diameter of the log?

2. Three people buy wood together. One pays the merchant 5 coins, another 3 coins, and the last 2 coins. It is found that in the transactions, 4 coins are left over. They wish to divide these proportionally among themselves. How many coins should each person receive?

3. A fox, a raccoon, and a hound pass through customs and together pay 111 coins. The hound says to the raccoon, and the raccoon says to the fox, "Since your fur is worth twice as much as mine, then the tax you pay should be twice as much!" How much should each pay?

4. If in one day, a person can make 30 arrows or fletch [put the feathers on] 20 arrows, how many arrows can this person both make and fletch in a day?

5. There are two piles, one containing 9 gold coins, the other 11 silver coins. The two piles have the same weight. One coin is taken from each pile and put in the other. It is now found that the pile of mainly gold weighs 13 units

less than the pile of mainly silver coins. Find the weight of a silver and a gold coin.

6. Now there is a wall 5 feet thick. Two rats tunnel from opposite sides. On the first day, the big rat tunnels 1 foot; the small rat also tunnels 1 foot. The big rat then doubles its daily rate, and the small rat halves its rate. Tell the number of days until the rats meet.

7. Given, a door and a measuring rod, both of unknown dimensions. Using the rod to measure the door, it is found that the rod is 4 feet longer than the width of the door, 2 feet longer than the height, and the same length as the diagonal. What are the dimensions of the door?

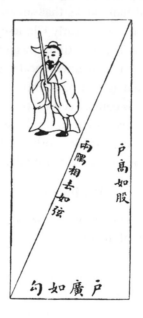

8. A two-door gate of unknown width is opened so that a 2-inch gap exists between the two doors. It is known that the open door's edge protrudes 1 foot from the door sill. What is the width of the gate?

9. One military horse cannot pull a load of 40 *dan*; neither can two ordinary horses, nor can three inferior horses. But one military horse and one ordinary horse can pull the load, as can two ordinary horses and one inferior horse, or three inferior and one military horse. How much can each horse pull?

10. Now a good horse and an inferior horse set out from Chang'an to Qi. Qi is 3000 *li* from Chang'an. The good horse travels 193 li on the first day and daily increases by 13 li; the inferior horse travels 97 li on the first day and

daily decreases by ½ li. The good horse reaches Qi first and turns back to meet the inferior horse. Tell: how many days until they meet, and how far has each traveled?

11. There is a circular pond inside a square field, and the area outside of the pond is 13.75 *mu*. It is known that the sum of the perimeter of the square and circle is 300 *bu*. Find the perimeters of the square and the circle. [1 mu = 240 bu^2.]

12. There is a four-sided field: the eastern side measures 35 paces; the western side, 45 paces; the southern, 25 paces; and the northern, 15 paces. Find the area of the field. Is the Chinese answer of 800 square paces correct?

13. Determine a number having remainders 2, 3, and 2 when divided by 3, 5, and 7, respectively.

14. Rabbits and pheasants are put in a basket. The top of the basket shows 35 heads, and the bottom shows a total of 94 feet. How many of each animal are in the basket?

15. Now there are six-headed four-legged animals and four-headed two-legged birds placed together. A count above the group gives 76 heads, and a count below gives 46 legs. Find the number of animals and birds.

16. Now there are three sisters who leave home together. The eldest returns once every 5 days, the second returns once every 4 days, and the youngest returns once every 3 days. Find the number of days before the three sisters meet together again.

17. A horse, halving its speed each day, travels 700 miles in 7 days. How far does it travel each day?

18. Now given: a guest on horseback rides 300 li in a day. The guest leaves his clothes behind. The host discovers them after ⅓ day, and he starts out with the clothes. As soon as he catches up with the guest, the host gives back the clothes and returns home in ¾ day. Assume the host rides without a stop. Tell: how far can he go in a day?

19. Now there was a woman washing bowls by the river. An official came by and asked, "Why are there so many bowls?" The woman replied, "There were guests in the house." The official then asked, "How many guests?" The woman said, "I do not know, but every two persons had a bowl of rice; every three persons had a bowl of soup; and every four persons had a bowl of meat. Sixty-five bowls were used altogether." How many guests were there?

20. A square walled city of unknown dimensions has four gates, one at the center of each side. A tree stands 20 bu from the north gate. One must

walk 14 bu southward from the south gate, and then turn west and walk 1775 bu, before he can see the tree. What are the dimensions of the city?

21. Four counties are required to furnish wagons to transport 250,000 *hu* of grain to a depot. There are 10,000 families in the first county, which is 8 days' travel to the depot; 9500 families in the second county, which is 10 days' travel; 12,350 families in the third county, 13 days' distant; and 12,200 families in the last county, which is 20 days' travel to the depot. The total number of wagons required is 1000. How many wagons are to be provided by each county according to the size of the population and the distance from the depot? [Let the number of wagons supplied be directly proportional to the population size and inversely proportional to the distance.]

22. Taking an excursion in the spring, I bring along a bottle of wine. On reaching a tavern, I double the bottle's contents and drink 1⁹⁄₁₀ *dou* in the tavern. After visiting four taverns, my bottle is empty. Permit me to ask, "How much wine was there at the beginning of the trip?"

23. There is a square pond, each side 10 feet long. Reeds grow vertically along the western bank and reach exactly 3 feet out of the water. On the eastern bank, another kind of reed grows exactly 1 foot out of the water. When the two reeds are made to meet, their tops are exactly level with the surface of the water. Permit me to ask how to determine these three things: the depth of the water and the length of each reed?

24. I have two reeds. One grows 3 feet and the other grows 1 foot on the first day. The growth of the first becomes every day half that of the preceding day, while the other grows twice as much as the day before. In how many days will the reeds be of equal height?

25. Two men start from the same point and begin walking in different directions. Their rates of travel are in the ratio of 7 to 3. The slower man walks towards the east. His companion walks to the south a distance of 10 bu and turns towards an intercepting path and proceeds until both men meet again. How many bu does each man walk?

26. Given: a wooden log with a diameter of 2 feet 5 inches from which a 7-inch-thick board is to be cut. What is the maximum possible width of the board?

27. A tree is 20 feet tall and has a circumference of 3 feet. There is a vine that grows at the bottom and winds equally seven times around the tree before it reaches the top. What is the length of vine?

28. A cow, a horse, and a goat were in a wheat field and ate some stalks of wheat. Damages of five baskets of grain were asked by the wheat field's owner. If the goat ate one-half the number of stalks as the horse, and the horse ate one-half of what was eaten by the cow, how much should be paid by the owners of the goat, horse, and cow, respectively?

29. In surveying a distant sea island, I erect two poles of the same length, 3 *zhang*, the distance between the poles being 1000 bu. Assume that the rear and front poles are aligned with the island. By moving back 123 bu from the front pole and observing the peak of the island from ground level, I notice that the tip of the front pole coincides exactly with the peak. Then I move 127 bu back from the rear pole and observe the island from ground level again. The top of the pole once again coincides with the peak. What is the height of the island? How far is it from the front pole?

[Note: 1 li = 300 bu = 180 zhang.]

30. An old Chinese general led his army to a river with a steep bank. Standing atop the bank, he held a stick 6 feet long perpendicular to himself. When the near end was 0.5 feet below his eye, he sighted the opposite side of the river over the far end of the stick. Without moving anything but his eyes, he sighted the near side of the river in line with a mark on the stick 2.6 feet from the near end. Having finished gazing, he lowered a rope down the bank and found the length to be 30 feet. What was the width of the river if the general's eye was 5 feet above the ground?

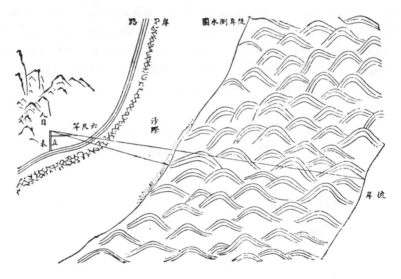

31. A bamboo shoot 10 *chi* tall has a break near the top. The configuration of the main stem and its broken portion formed a triangle. The top touches the ground 3 chi from the stem. What is the length of the stem remaining erect?

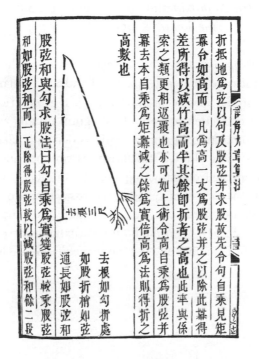

32. In the center of a square pond whose side is 10 chi grows a reed whose top reaches 1 chi above the water level. If we pull the reed towards the bank, its top is even with the water's surface. What is the depth of the pond and the length of the plant?

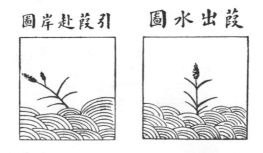

WHAT ARE THEY DOING?

Using Geometric Approximation (Completing the Square) to Obtain the Square Root of a Number

The *Great Encyclopedia of the Yongle Reign*, written in the fifteenth century, contains the pages shown in figures 6.1 and 6.2. To many readers, the diagram in figure 6.1 may seem familiar: it is a geometric derivation of the square root extraction process, the diagram usually associated with the process of "completing the square." This section of the *Encyclopedia* records the work of Yang Hui (ca. 1270), who is commenting on the traditional Chinese method of extracting roots of numbers and polynomial equations. From the early Han dynasty onwards, Chinese mathematicians, using their counting boards, could extract square and cube roots of numbers with great accuracy. Their methods were extended to obtaining solutions for quadratic equations and polynomial equations of higher degree.

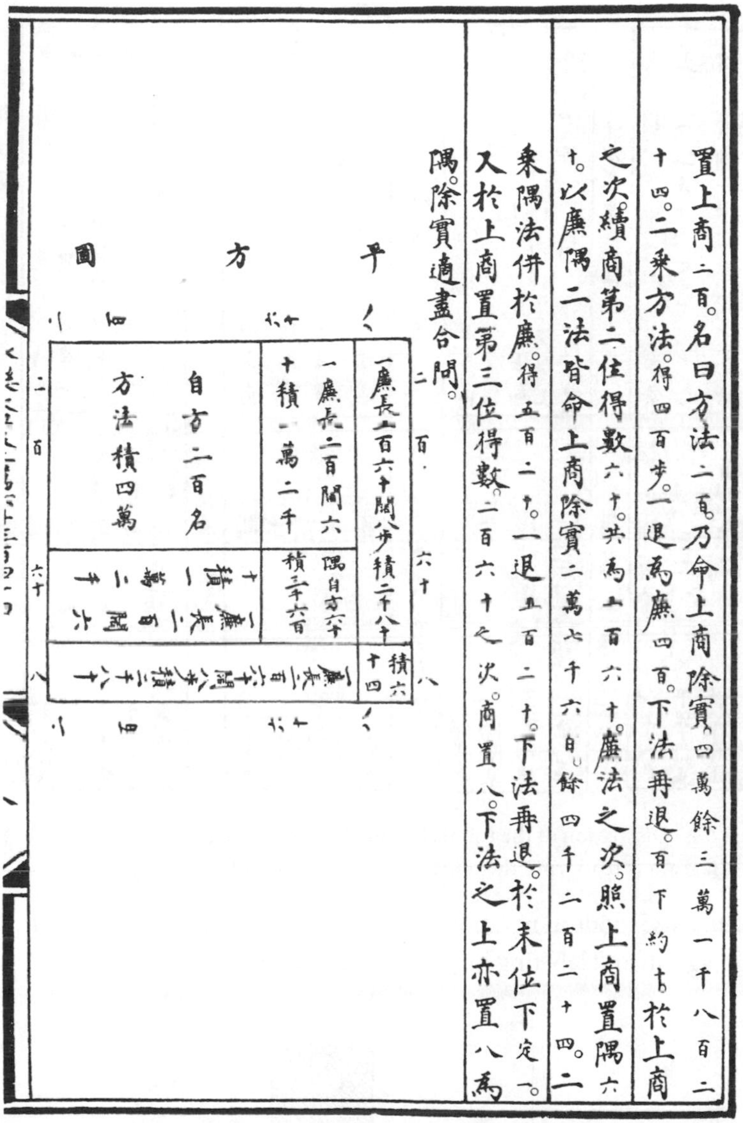

置上商二百。名曰方法。二
十四。二乗方法。得四百步。一退為廉四百。下法再退。百下約亦於上商
之次。續商第二位得數六十。共為二百六十。廉法之次。照上商置隅六
十。以廉隅二法皆命上商除實二萬七千六百。餘四千二百二十二
乗隅法併於廉。得五百二十。一退五百二十。下法再退於末位下定一。
又於上商置第三位得數二百六十之次。商置八下法之上亦置八為
隅除實適盡合問。

圖　方　平

方法積四萬
自方二百名
一廉長二百闊六
十積一萬二千
一廉長二百六十闊八步積二千八十
隅自乗八
積六十四

二百　六十　八
四
二百　六十　八

Figure 6.1. Fifteenth-century Chinese illustration for the process of "completing the square."

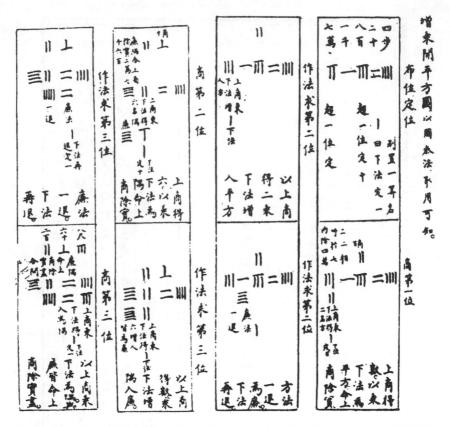

Figure 6.2. Discussion of the rod technique for computing the square root of 71,824. (The diagram is read from the upper right to the lower left.)

Let us use modern notation and explore what is happening. Redrawing this diagram and labeling its parts using modern notation:

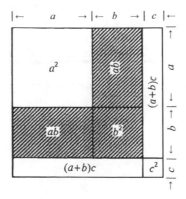

Then the area of the square is given by $(a+b+c)^2 = a^2 + b^2 + c^2 + 2(ab+ac+bc)$.

If given a number N, let $N = (a+b+c)^2$ and then $\sqrt{N} = a+b+c$. Thus, to approximate a square root of a number, N, take the following steps:

1. Choose a value for $a \Rightarrow a^2$.
2. Then compute $b+c \leq (N-a^2)/2a$.
3. Choose b accordingly.
4. Then choose c, where $c \leq \dfrac{N-(a+b)^2}{2(a+b)}$.
5. Then $\sqrt{N} = a+b+c$.

Figure 6.3 relates the computing directions given by Yang to solve a specific problem from the *Jiuzhang suanshu*. Following his directions and using the notation employed above, we are guided in solving the stated problem:

Given a square of area 71,824 square *bu*. What is the side of this square?

1. Find $\sqrt{71,824}$
2. Approximate $\sqrt{70,000} \approx 200$; then $b+c \leq 31,824/400 \approx 79$.
3. Since b is a 10s digit, we try 70 for b, but this is too large. There-fore, let $b = 60$.
4. Lastly, we approximate c using $c \leq (71,824 - 260)^2/(2 \times 270) = 7.82 \approx 8$.
5. We find $\sqrt{71,824}$ is $200 + 60 + 8 = 268$.

7   India

Eighteenth-century Indian astronomers search the heavens. The instrument they are using is a theodolite, which can determine vertical and horizontal distances. While really concerned with astrology, they were devising principles of mathematical astronomy.

In ancient India records were kept on palm leaf sheets. This material was very fragile and did not survive for a long time. Therefore, few works on old Indian or Hindu mathematics exist today. The works that do exist are written in Sanskrit, a difficult language to translate. The earliest known Hindu mathematical works, the Sulbasutras (cord-stretching manuals) were written before the Christian era. These are handbooks for priests and describe the geometric construction of altars and ritual techniques used in astronomy. The Bakhshali Manuscript (ca. 400 CE) provides examples of simple arithmetic. Later mathematicians such as Brahmagupta (ca. 628) and Bhāskara II (1150) did compile collections of problems. These problems are often written in a poetic or fanciful form as an aid to memorization. Some of these problems are given below.

Problems

1. Find a number having remainder 29 when divided by 30 and remainder 3 when divided by 4.

2. One person possesses 7 asavas horses, another 9 hayas horses, and another 10 camels. Each gives 2 animals away, one to each of the others. Now the three men are equally well off. Find the value for each animal and the total value of the livestock possessed by each person.

3. On an expedition to seize his enemy's elephants, a king marched 2 *yojanas* the first day. Say, intelligent calculator, with what increasing rate of daily march did he proceed, since he reached his foes' city, a distance of 80 yojanas, in a week?

4. Twenty-three weary travelers entered a delightful forest. There they found 63 numerically equal piles of plantain fruit. Together, they gathered 7 more plantains; then they divided all the plantains among themselves in a manner such that no plantains remained. For this to happen, what was the smallest number of plantains possible in each pile?

5. The third part of a necklace of pearls, broken in a lover's quarrel, fell to the ground; its fifth part rested on the couch; the sixth part was saved by the wench; and the tenth part was taken by her lover. Six pearls remained. How many pearls composed the necklace?

6. A fish is resting at the northeast corner of a rectangular pool. A heron standing at the northwest corner spies the fish. When the fish sees the heron looking at him, he quickly swims towards the south. When he reaches the south side of the pool, he has the unwelcome surprise of meeting the heron,

who has calmly walked due south along the side and turned at the southwest corner of the pool and proceeded due east, to arrive simultaneously with the fish on the south side. Given that the pool measures 12 units by 6 units, and that the heron walks as quickly as the fish swims, find the distance the fish swam.

7. In a certain lake swarming with red geese, the tip of a lotus bud was seen to extend a span [9 inches] above the surface of the water. Forced by the wind, it gradually advanced and was submerged at a distance of 2 cubits [40 inches]. Compute quickly, mathematician, the depth of the pond.

8. Friend, tell me quickly the square of $3\frac{1}{2}$, and then the square root of the square, and the cube root of it, and then the root of the cube, if you know fractional squares and cubes.

9. A traveler on a pilgrimage gave $\frac{1}{2}$ of his money at Allahabad, $\frac{2}{9}$ of the rest at Benares, $\frac{1}{4}$ of the remainder in toll fees, and $\frac{6}{10}$ of the remainder at Patna. After this, 63 gold coins were left over, and he returned to his home. Tell me the initial amount of money.

10. In a triangle, when the two sides are 10 and 17 and the base is 9, tell me quickly, mathematician, the altitude and also the area of the triangle.

11. Of a collection of mango fruits, the king took 16; the queen $\frac{1}{5}$ of the remainder; the three princes, $\frac{1}{4}$, $\frac{1}{3}$, and $\frac{1}{2}$ of the successive remainders; and the youngest child took the remaining 3 mangoes. Oh you who are clever in miscellaneous problems on fractions, give out the measure of that collection of mangoes.

12. The eighth part of a troop of monkeys, squared, were skipping in a grove and delighted with their sport. Twelve remaining monkeys were seen on the hill, amused with chattering to each other. How many were there in all?

13. Given: a vertical pole of height 12 feet. The ingenious man who can compute the length of the pole's shadow, the difference of which is known to be 19 feet, and the difference of the hypotenuse formed, 13 feet, I take to be thoroughly acquainted with the whole of algebra, as well as arithmetic.

14. A powerful, unvanquished, excellent black snake that is 80 *angulas* in length enters into a hole at the rate of $7\frac{1}{2}$ angulas in $\frac{5}{14}$ of a day, and in the course of a day, its tail grows $11\frac{1}{4}$ angulas. O ornament of arithmeticians, tell me what time this serpent fully enters its hole.

15. The mixed price of 9 citrons and 7 fragrant wood apples is 107; again, the mixed price of 7 citrons and 9 fragrant wood apples is 101. O you arith-

metician, tell me quickly the price of the citron and the wood apple here, having distinctly separated those prices well.

16. The square root of half the number of bees in a swarm has flown out upon a jasmine bush; $8/9$ of the swarm has remained behind. A female bee flies about a male bee that is buzzing within a lotus flower into which he was lured in the night by its sweet odor, but in which he is now is imprisoned. Tell me, most enchanting lady, the number of bees.

17. A quantity when divided by 12 leaves a remainder of 5, and furthermore it is seen to have a remainder of 7 when divided by 31. What should such a quantity be?

18. The sales tax on garments is $1/20$ of their value. A certain man buys 42 garments, paying in *panas* (copper coins). Two garments are taken away as tax; also,10 panas are paid as the remaining part of the tax. What is the price of a garment, O learned one?

19. There is a hole at the foot of a pillar 9 *hastas* high, and a pet peacock standing on top of it. Seeing a snake returning to its hole at a distance from the pillar equal to three times its height, the peacock swoops down upon the snake slantwise. Say quickly, how far from the pole does the meeting of their paths occur?

20. One monkey came down a tree of height 100 and went to a pond a distance of 200. Another monkey, leaping some distance above the tree, went diagonally to the same place. If their total distances traveled are equal, tell me quickly, learned one—if you have a thorough understanding of calculation— how much is the height of the leap?

21. In a triangular figure, when the base is equal to 14 and the arms 13 and 15, tell me quickly the altitude and the two base segments, and also the amount of equal parts known as the area.

22. One person has 300 coins and 6 horses; another has 10 horses having an equal price and has a debt of 100 coins. The two have equal wealth. What is the price of a horse?

23. The 8 rubies, 10 emeralds, and 100 pearls that are in your ear ornament were purchased by me for you at an equal price. The sum of the prices of this triad of gems was half 100 diminished by 3. Tell me, my dear, the price of each separately if, O auspicious one, you are clever in mathematics.

24. The shadows of two equal gnomons are observed to be respectively 10 and 16 *angulas*, and the distance between the tips of the shadows is seen as 30. Both the upright side and the base should be found.

25. Let the earth be 14 and the face 4 units. The two chief ears should measure 13; tell the length of the lines that intersect at the top and the area.

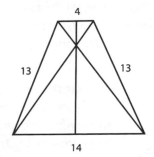

26. A fifth part of a troop of monkeys, minus three, squared has gone to a cave: one is seen having climbed to a branch of a tree. Tell, how many are there?

27. Arjuna, exasperated in combat, shot a quiver of arrows to slay his foe, Carna. With half his arrows he parried those of his antagonist; with four times the square root of the quiverful he killed his horse; with six arrows he slew Carna's charioteer; with three he demolished the umbrella, standard, and bow; and with one he cut off the head of his foe. How many arrows did Arjuna disperse?

WHAT ARE THEY DOING?
Applying the Rule of Three

One of the basic rules of ancient Indian mathematics was *trairasica*, or the rule of three. The mathematician Brahmagupta (ca. 628) explained the rule as follows:

> In the rule of three, argument, fruit, and requisition or desire are the names of the terms. The first and last terms must be similar. Requisition multiplied by fruit and divided by the argument or measure result in the produce or fruit of the desire. (Smith 1958, 2:483)

A typical problem on which this rule will be applied is the following:

> If 1¼ *pala* of sandalwood costs 10½ *pana*, how much does 9¼ *pala* of sandalwood cost?

A mathematical scribe would multiply the fruit by the desire and divide the result by the measure to obtain the fruit of desire, 77⁷⁄₁₀ *pana*.

Today, we recognize this procedure as the expression of a simple proportion: $a/b = c/x$ and $x = bc/a$.

The rule of three was adopted by Arab merchants and introduced into Europe. Among the Italian merchants it became known as *regula del tres*. Other names were also attached to it—the golden rule, indicating its importance, and the merchants' rule, reflecting its principal use. Indeed, this simple mathematical technique was highly valued in the merchant community. Hodder in his arithmetic book of 1683 comments on the rule of three as follows:

> The Rule of Three is commonly called the Golden Rule, and indeed, it might be so determined: whereas as gold transcends all other metals, so does this rule to all others in arithmetic. (p. 87)

Similar rules based on proportions would also be developed such as the rule of five and the rule of seven.

8 Islam

"Arab" surveyors take measurements, as depicted in a fifteenth-century woodcut print in a commentary on Cicero's *Somnium Scipionis*. The term "Arab" referred to any Islamic scholar at this time, since all scholarly work done in the Islamic community was written in Arabic.

The religion of Islam, founded by the prophet Muhammad, dates to the year 622. Originating on the Arabian peninsula, Islam united diverse tribes and peoples into one group obeying the same religious rules and following the same associated cultural rituals, and eventually spread worldwide. In the year 750, the caliph al-Mansur established a capital in the city of Baghdad. The city became a center for Islamic learning. Ancient Greek classics were collected and translated into Arabic. Scholars studied the mathematics and science of the Greeks.

In 766, a formal research institution, the *Bayt al-Hikma*, or House of Wisdom, was founded in Baghdad. It attracted scholars from around the Islamic world. Some of these scholars began to write their own books on mathematics. Perhaps the most influential of these books is one written by Muhammad ibn Musa al-Khwarizmi (ca. 780–850) entitled *Kitab al-jabr wa'l muqabaluh*. This book discusses what we would know as the solving of equations by using two processes: "balancing" and "restoring." In Arabic the word for "restoring" is *al-jabr*. When this book was translated and brought to the West, it became a book of algebra.

When we speak of "Islamic mathematics," we are referring to the mathematical developments that took place during the Abbasid caliphate (750–1000), when an empire founded on the principles of Islam ruled much of the Middle East, North Africa, and parts of Europe. Contributors to this mathematics were not necessarily Muslims. The freedom of scholarship that existed in all parts of the empire allowed Christians, Jews, and other non-Muslims to live and work. The problems that follow are typical of those found in the corpus of Islamic mathematics from this period.

Problems

1. One says that 10 is divided into three parts, and if the small part is multiplied by itself and added to the middle one multiplied by itself, the result is the large one multiplied by itself, and when the small part is multiplied by the large part, it equals the middle part multiplied by itself. Find the parts.

2. The number 50 is divided by a certain number. If the divisor is increased by 3, the quotient decreases by 3.75. What is the number?

3. I have divided 10 into two parts, and have divided the first by the second, and the second by the first and the sum of the quotients is $2\frac{1}{6}$. Find the parts.

4. Suppose 10 is divided into two parts, and the product of one part by itself equals the product of the other part by the square root of 10. Find the parts.

5. Suppose 10 is divided into two parts, each one of which is divided by the other, and when each of the quotients is multiplied by itself and the smaller is subtracted from the larger, there remains 2. Find the parts.

6. One says that 10 garments were purchased by two men at a price of 72 *dirhams*. Each pays 36 dirhams. The garments vary in value. The price of each garment of one man is 3 dirhams more than the price for each garment of the other. How many garments did each man buy?

7. Given a number, take one-third of the number away from itself and add 2. If the result is multiplied by itself, it equals the number plus 24. What is the number?

8. Given a triangular piece of land having two sides 10 yards in length and a base of 12 yards, what is the largest square that can be constructed within this piece of land so that one of its sides lies along the base of the triangle?

9. You have two sums of money, the difference of which is 2 dirhams; you divide the smaller sum by the larger, and the resulting quotient is equal to $\frac{1}{2}$. What are the two sums of money?

10. The number of 50 is divided by a certain number. If the divisor is increased by 3, the quotient decreases by 3.75. What is the number?

11. A woman dies, leaving her husband, a son, and three daughters. Calculate the fraction of her estate each will receive.

[Note: The conditions of Islamic law must be followed—that is, the husband must receive one-fourth share, and his son twice as much as a daughter.]

12. A woman dies, leaving her husband, a son, and three daughters, but she also leaves to a stranger $\frac{1}{7} + \frac{1}{8}$ of her estate. Calculate the shares for each heir.

[Note: by Islamic law, a legacy outside the family cannot exceed one-third the estate unless the natural heirs agree; the family members would then share the estate according to the conditions noted above after the external legacy is paid.]

13. Given: a triangle with sides of lengths 14, 13, and 15. Using the side of length 13 as the base of the triangle, find the length of its altitude.

14. I have divided 10 into two parts. I have afterwards divided one by the other and the quotient I obtained is 4. What are the two parts?

15. Two squares and 10 roots are equal to 48 dirhams. What is the root?

16. I have multiplied one-third of a thing and one, by one-fourth of the thing and one, and the product was 20. What is the value of the thing?

17. I have divided 10 into two parts; I have divided each of them by itself, and when I added the products together, the sum was 58 dirhams. What were the parts?

18. I have multiplied one-third of a root by one-fourth of the root, and the product is equal to the root plus 24. What is the root?

19. There is a square. Two-thirds of one-fifth of it are equal to one-seventh of its root. What is its root?

20. If someone says a workman receives a pay of 10 dirhams per month, how much must he be paid for six days?

21. One says that 10 is divided into two parts. One part is multiplied by itself and the other by the root of 8. Then if one subtracts the quantity of the product of one part times the root of 8 from the product of the other part multiplied by itself, it gives 40. What are the two parts?

22. One says that 10 is divided into two parts, each of which is divided by the other, and when each of the quotients is multiplied by itself, the smaller is subtracted from the larger, then there remains 2. What were the parts?

23. The mathematician ibn al-Haytham, known in the West as Alhazen (965–1040), devised a method of constructing odd-order magic squares. His method for constructing a fifth-order square is demonstrated below:

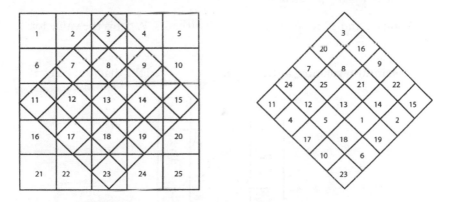

The oblique square on the right is rotated 45 degrees counterclockwise. Use this method to construct a seventh-order magic square.

24. The Persian mathematician Abu Bakr al-Karaji (d. 1019) devised a construction whereby, given a circle, one could construct another circle with area $1/n$ of the given circle. The construction is shown in the diagram below. Prove that it is correct.

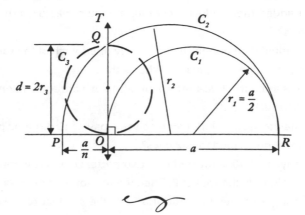

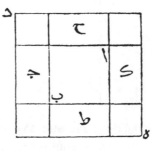

WHAT ARE THEY DOING?

Solving a Quadratic Equation by Completing the Square

Al-Khwarizmi (ca. 825) in his *Algebra* demonstrated how to solve six types of equations. The solution for one type of equation was explained with the following problem:

A square and 10 roots are equal to 39 dirhams. What is the value for a root?

In explaining a solution, al-Khowarizmi employed the diagram below:

84

In our analysis of the solution, let us use modern notation. We are solving the problem $x^2 + 10x = 39$. First, we construct a square with side x and then extend its sides to enclose rectangles with sides measuring $10/4 = 2\frac{1}{2}$.

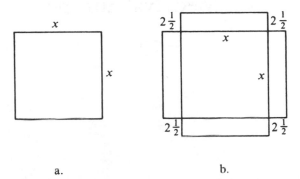

a.   b.

The figure on the right now geometrically represents $x^2 + 10x$, the left-hand side of the given equation. We "complete this square" by adding four small squares of area $(5/2)^2$ to its corners. To balance the situation, we must add $25/4 \times 4$ to the other side of the equation. $39 + 25 = 64$

Thus, the area of the completed square is 64 and its side is 8. This makes the side of the smaller square (the original square) $8 - 5 = 3$; therefore, $x = 3$.

As we have seen from the previous problems, many early peoples employed a geometric approach to algebraic operations. With a product of two quantities conceived of as the area of a rectangle, the series of mathematical operations that followed could be visually justified. Early algebra was based on this use of visualization. The ancient Babylonians employed a method of "completing the square" to solve quadratic equations, as did the Chinese in their perfection of root extraction techniques. So too, when al-Khwarizmi compiled his algebra, many of his techniques were based on the geometric method of completing the square.

9 Medieval Europe

During the Middle Ages, the works of Pythagoras were held in high esteem. An old woodcut print from a 1491 arithmetic depicts Pythagoras teaching youths. Note the mathematical discrepancies in this illustration.

Pictagoras aritbmetrice introductor

With the fall of the Roman Empire in about 500 CE, Europe reverted to a more primitive state. The system of taxation collapsed, and the monetary system was not maintained. Road networks deteriorated, and transportation became difficult. Villages became isolated. A feudal system arose, whereby the agriculture-based population was ruled over and protected by local aristocratic warlords. In turn, the peasants' products and services belonged to their lord. The only unifying institutional influence at this time was that of the Catholic Church. The church was mainly interested in spiritual matters; the pursuit and promotion of the sciences, including mathematics, was of minor importance.

During this time, only two major works of mathematics appeared: *Propositiones ad acuendos juvenes* (800) and *Liber Abaci* (1202). Charlemagne, a Frankish warlord who became emperor of the part of Europe known as the Holy Roman Empire in the year 800, established a court school in the 780s and brought a monk, Alcuin of York, from England to teach at the school. Alcuin wrote the *Propositiones*, or *Problems to Sharpen the Young*, a collection of 56 mathematical puzzles intended for the training of court pages.

By the tenth century the increased tempo of mercantile and artisan activities in Europe supported international trade. European merchants traveled far in search of exotic goods and profit. Bankers, money lenders, and traders needed mathematics for their record keeping. *Liber Abaci* (1202), or *The Book of Computation*—written by the Italian merchant Leonardo of Pisa (ca. 1175–1250), popularly known as Fibonacci—taught the techniques of arithmetic and simple algebra. This was one of the first works to introduce Hindu Arabic numerals and the use of common fractions into Italy. Fibonacci learned much of his mathematics from Arab teachers while living in North Africa.

Problems

1. Three hundred pigs are to be prepared for a feast. They are to be prepared in three batches on three successive days with an odd number of pigs in each batch. How can this be accomplished?

2. There is a lion in a well whose depth is 50 palms. He climbs $\frac{1}{7}$ of a palm daily and slips back $\frac{1}{9}$ of a palm. In how many days will he get out of the well?

3. There is a tree with 100 branches; each branch has 100 nests; each nest, 100 eggs; each egg, 100 birds. How many nests, eggs and birds are there?

4. On a certain ground stand two poles 12 feet apart; the lesser pole is 35 feet in height and the greater 40 feet. It is sought, if the greater pole will lean on the lesser, then in what part will it touch?

5. A certain man had in his trade four weights with which he could weigh integral pounds from 1 up to 40. How many pounds was each weight?

6. A leech invited a slug for lunch a *leuca* away. But he could only crawl an inch a day. How long will it take the slug to get his meal? [1 leuca = 1500 paces; 1 pace = 5 feet.]

7. A four-sided town measures 1100 feet on one side and 1000 feet on the other side; on one edge, 600, and on the other edge, 600. I want to cover it with roofs of houses, each of which is to be 40 feet long and 30 feet wide. How many dwellings can I make there?

Alcuin's answer is 520 houses. Is this correct?

8. Two walkers saw some storks and wondered how many there were. Conferring, they decided: if there were the same number again, and again, and then half of a third of the sum that would make 2 more, that would be 100. How many storks were seen?

9. A merchant wanted to buy 100 pigs for 100 pence. For a boar, he would pay 10 pence and for a sow 5 pence, while you would pay 1 penny for a couple of piglets. How many boars, sows, and piglets must there have been for him to have paid exactly 100 pence for 100 animals?

10. A dish weighing 30 pounds is made of gold, silver, brass, and lead. It contains three times as much silver as gold, three times as much brass as silver, and three times as much lead as brass. How much is there of each metal?

11. A cask is filled with 100 *metretae* through three pipes. One-third plus a sixth of the capacity flows in through one pipe, one-third of the capacity flows in through another pipe, but only one-sixth of the capacity flows in through the third pipe. How many *sextarii* flow through each pipe? [1 metreta ≈ 9 gallons; 1 sextarius ≈ 1 pint; 1 metreta = 72 sextarii.]

12. I have a cloak 100 cubits long and 80 cubits wide. I wish to make small cloaks with it; each small cloak is 5 cubits long, and 4 wide. How many small cloaks can I make?

13. A father, when dying, gave to his sons 30 glass flasks, of which 10 were full of oil, 10 were half full, and the last 10 were empty. Divide the oil

and the flasks so that each of the three sons received equally of both glass and oil.

14. A king ordered his servants to collect an army from 30 manors in such a way that from each manor he would take the same number of men he had collected up until then. The servant went to the first manor alone; to the second he went with one other; to the next he took three with him. How many were collected from the 30 manors?

15. An ox plows the field all day. How many footprints does he leave in the last furrow?

16. Two men were leading oxen along a road, and one said to the other, "Give me two oxen, and I will have as many as you have." Then the other said, "Now you give me two oxen, and I'll have double the number you have." How many oxen were there, and how many did each have?

17. Three friends, each with a sister, needed to cross a river. Each one of them coveted the sister of another. At the river, they found only a small boat, in which only two of them could cross at a time. How did they cross the river without any of the women being defiled by the men?

18. A man and a woman, each the weight of a cartload, with two children, who together weigh as much as a cartload, have to cross a river. They find a boat which can take only one cartload. Make the transfer if you can without sinking the boat.

19. There is a field which is 200 feet long and 100 feet wide. I want to put sheep in it so that each sheep has a space of 5 feet × 4 feet. How many sheep can I put in there?

20. A four-sided field measures 30 *perches* down one side and 32 down the other; it is 34 perches across the top and 32 across the bottom. How many acres are included in this field?

21. There is a wine cellar 100 feet long and 64 feet wide. How many casks would it hold if each cask is 7 feet long and 4 feet wide across the middle, and there is to be one path 4 feet wide?

22. A gentleman has a household of 30 people and orders that they be given 30 measures of grain. He directs that each man should receive 3 measures, each woman 2 measures, and each child ½ measure. How many men, women, and children are there?

23. A man wanting to build a house contracted with six builders, five of whom were master builders, and the sixth an apprentice, to build it for him.

He agreed to pay them a total of 25 pence a day, with the apprentice to get half the rate of a master builder. How much did each receive per day?

24. A man in the east wanted to buy 100 assorted animals for 100 shillings. He ordered his servant to pay 5 shillings for a camel, 1 shilling for an ass, and 1 shilling for 20 sheep. How many camels, asses, and sheep did he buy?

25. A Barbary cane that is 8 palms long is sold for $4^7/_{10}$ *bezants*. It is sought how much $2^1/_4$ palms are worth?

26. Two men make a company in which one puts in 15 pounds, 7 *soldi*—that is, $15^7/_{20}$ pounds—and the other puts in 19 pounds. Their profit together makes 14 pounds, 14 soldi, and 5 *denari*, or 14 and $^5/_{12}$ and $^{14}/_{20}$ pounds. How much profit should each receive?

27. A certain man receives 7 bezants in a month for his labor, and if some of the time he does not labor, he pays back 4 bezants by a monthly rule. He stays for a month; sometimes he labors, sometimes he does not; thus, he has 1 bezant for when he labored, discounting when he did not labor. It is sought how much in the month he labored and how much not. [Assume 1 month = 30 days.]

28. A certain man buys in Constantinople 90 *modia* of corn, millet, beans, barley, and lentils for $21^1/_4$ bezants. Now, 100 modia of corn is sold for 29 bezants, barley truly for 25 bezants, but millet for 22 bezants and beans for 18 bezants. Lentils cost 16 bezants. It is sought how much he buys of each grain.

29. Two men having bezants find two horses for sale, the second of which is worth 2 bezants more than the price of the first. The first man with his bezants, and having one-third of the second man's bezants, proposes to buy the first horse. Truly, the second man, having one-fourth of the first's bezants, proposes to buy the second horse. All of these are made up of integral numbers. The price of each horse is sought and how many bezants each man has.

While almost all problems at this time were written in Latin, some Jewish scholars wrote in Hebrew. Levi ben Gershon (1288–1344)—a rabbi, philosopher, and mathematician—was such a scholar. He lived in a Jewish community in southern France and wrote several books on mathematics. The following problems are from his *Maaseh hoshev* (The art of calculation), written in 1321.

30. Given, $^2/_5 + ^3/_4 + ^1/_3$ of a number equals 20. May I ask the value of the whole number?

31. Given, $3/7 + 4/5$ of an unknown number exceeds $3/2 + 1/4$ of the unknown number by 20. What is this number?

32. The cost of 11 measures of wheat is 17 *dinars*. What is the cost of 15 measures of this wheat?

33. A barrel has various holes in it. The first hole empties the full barrel in 3 days; the second hole empties the full barrel in 5 days; another hole in 20 hours; and the remaining hole empties a full barrel in 12 hours. With all the holes open, how long will it take the barrel to empty?

34. A merchant sells two drugs. The price of the first is 17 *peshuts* per liter and the price of the second is 24 peshuts per liter. A buyer wants to buy a mixture of the two at 19 peshuts. What fraction of each of these drugs will make up the required measure?

35. Reuven bought $2/5$ and $3/7$ of a measure for $7^0/11$ dinars. He sold 4/9 of the measure for $8^3/7$ dinars. In all, he invested 100 dinars. Did he lose or gain on this transaction? How much?

36. The sum of two numbers is 13, and their product is 17. What are the numbers?

37. A number plus $2/7$ and $1/9$ of a second number makes 20. If you add the second number plus $2/5$ the first number, you also get 20. What are the numbers?

38. Given two numbers, when you add the first number to the second number, its ratio to the third number equals $3^2/5 + 1/7$. When the first is added to the third, its ratio to the second equals $7^2/3 = 1/4$. The second number 30. We want to know what is the value of each of the remaining numbers.

WHAT ARE THEY DOING?
Early Attempts at Graphing

Because medieval Europe lacked standard mathematical notation, authors at this time used various visual schemes to represent mathematical relationships. They employed sketches and pictures and, in some cases, figures that we would call "mathematical doodles" to illustrate and express some of their ideas. Figure 9.1 shows a hand-drawn illustration from a tenth-century manuscript depicting the positions of the "seven heavenly

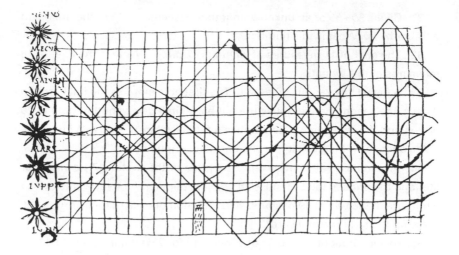

Figure 9.1. Early "graph" contained in Codex Latinus 14436, Bayerische Staats-Bibliothek, Munich.

wanderers"—that is, the five planets, the sun, and the moon—in the night sky during the period of one month.

Examine this diagram carefully. Can you identify the names of the heavenly wanderers given at the left? Is the grid of boxes a coordinate system? If so, what do they represent? What do the paths drawn across this grid represent?

In the fourteenth century, the Parisian scholar Nicole Oresme made a study of bodies moving through space. In about 1350, Oresme published *Tractatus de configurationibus qualitatum et motuum,* a work on the configurations of qualities in which he attempted to use geometric concepts to depict "the extension and intensity of any quality." Figure 9.2 shows a page in a later edition of his work and shows his marginal explanatory sketches.

If we interpret these diagrams as representing the velocity of a moving body over time, where the subdivisions of the baselines represent equal intervals of time, then the area enclosed by these curves represents distance traveled over a period of time (velocity × time = distance). Area is seen to represent a physical quantity, distance.

It would be such experimentation with expressions to represent mathematical concepts and interactions that provided us with the system of mathematical notation we know and use today.

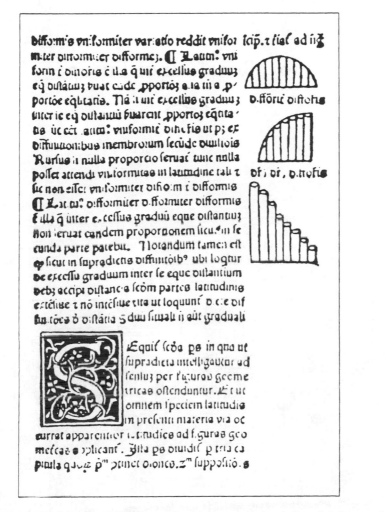

Figure 9.2. Page from Oresme's *Tractatus* with marginal sketches explaining possible variations of distance traveled over a period of time.

10 Renaissance Europe

An anonymous woodcut print from fifteenth-century Strasbourg shows a merchant calculating the exchange of money employing a counting table. Two curious burghers discuss the transaction.

With the appearance of Fibonacci's *Liber Abaci* in 1202, a new type of mathematics book appeared in Europe. This book was intended for merchants; it promoted a use of the Hindu Arabic numerals, discussed the particular topics of interest to merchants, and allowed for computation using a pencil and paper rather than the counting table. The new mathematics books that followed Fibonacci's work were sometimes called *Practica* because of their emphasis on practical, applied mathematics. Initially, they appeared in manuscript form. With the advent of printing, they multiplied quickly and became very popular. The first printed arithmetic book, the *Treviso Arithmetic*, appeared in Italy in 1478 and is named after the town in which it was published. It stressed all the mathematics needed by merchants: exchanging money, computing discounts, forming partnerships, and imposing interest payments on money. The first printed arithmetic in Germany appeared in 1482; in France and Spain, in 1512; in Portugal, in 1519; and in England, in 1537.

Thus, from the thirteenth century through the sixteenth, an extensive literature on using mathematics appeared in Europe. The following problems represent the mathematical concerns of this time and also tell us something about daily life. Diverse and strange monetary units and units of measure for distance and weight may appear in some of these problems and can be translated into modern terms if necessary.

Problems

1. Three circles are mutually tangent. The radii of one circle are 10 units, the radii of the second are 15 units, and the radii of the third circle are unknown. If the centers of these circles are connected by straight lines, the area of the enclosed triangle is 340.74 square units. What are the radii of the third circle?

2. Find two numbers, x and y, such that their sum is 10 and $x/y + y/x = 25$.

3. A man drinks up a barrel of wine in 14 days. If he drinks together with his wife, it will be empty in 10 days. In how many days will his wife drink up the barrel alone?

4. Three men have a pile of money, their shares being $1/2$, $1/3$, and $1/6$. Each man takes some money from the pile until nothing is left. The first man then returns $1/2$ of what he took, the second $1/3$, and the third $1/6$. When the total as returned is divided equally among the men, it is found that each receives

what he was originally entitled to. How much money was in the original pile, and how much did each take?

5. Two wine merchants enter Paris, one of them with 64 casks of wine, the other with 20. Since they do not have enough money to pay the custom duties, the first pays 5 casks of wine and 40 francs, and the second pays 2 casks of wine and receives 40 francs in change. What is the price of each cask of wine and the duty on it?

6. The French mathematician Nicolas Chuquet (fl. 1484) claimed that if given positive numbers a, b, c, and d, then $(a+b)/(c+d)$ lies between a/c and b/d. Is he correct? Prove your answer.

7. A merchant bought 50,000 pounds of pepper in Portugal for 10,000 *scudi* and paid a tax of 500 scudi. He transported the pepper to Italy at a cost of 300 scudi and paid another duty of 200 scudi. The next transport, to Florence, cost 100 scudi, and he had to pay an import duty of 100 scudi to that city. Lastly, the government demanded a tax from each merchant of 1000 scudi. Now he is perplexed to know what price to charge per pound for his pepper so that, after all these expenses, he may earn a profit of $1/10$ scudi a pound.

8. Leonardo da Vinci posed this problem and the next one in 1505. Given, a pendulum as shown on the left. The height of the pendulum is 2 units, and its horizontal width is 2 units. What is the area of the pendulum?

9. Given, the cat's eye at the left. Let the radius of the eye be given by R. What is the area of the pupil?

10. If I were to give 7 pennies to each beggar at my door, I would have 24 pennies left in my purse. I lack 32 pennies to be able to give each 9 pennies apiece. How many beggars are there, and how much money do I have?

11. In order to encourage his son in the study of arithmetic, a father agreed to pay him 8 pennies for every problem solved correctly and to charge him 5 pennies for each incorrect solution. After the son completed 26 problems,

neither owed anything to the other. How many problems did the son solve correctly?

12. There is a number which when divided by 2, or 3, or 4, or 5, or 6 always has a remainder of 1 and is truly divisible by 7. It is sought what is the number.

13. One of two men had 12 fish, and the other had 13 fish, and all of the fish were of the same price. From the first man, a customs agent took away 1 fish and 12 denarii for payment. And from the other he took 2 fish and gave him back 7 denarii. Find the customs fee and the price of each fish.

14. In a right triangle, let the perpendicular be 5 and the sum of the base and hypotenuse 25. Find the lengths of the base and hypotenuse.

15. In a right triangle, the hypotenuse is 9.434 and the sum of the sides around the right angle is 13. Find the lengths of the sides around the right angle.

16. The square of a certain number multiplied by itself and by 200 is 446,976. What is the number?

17. One hundred men besieged in a castle have sufficient food to allow each of them bread to the weight of 14 lots a day for 10 months. Seven months and 20 days later, the men are warned that the castle can receive no help for 4 months longer. How much bread should each man be allotted, counting each month as 30 days?

18. A certain slave fled from Milan to Naples, going 1/10 of the whole journey each day. At the beginning of the third day, his master sent a slave after him, and this slave went 1/7 of the whole journey each day. I do not know how far it is from Milan to Naples, but I wish to know when the latter overtook him.

19. Four men having denarii found a purse of denarii. The first man said that if he would have the denarii of the purse, then he would have twice as many as the second; the second, if he would have the purse, then he would have three times as many as the third; the third, if he would have it, then he would have four times as many as the fourth; and the fourth, five times as many as the first. It is sought how many denarii each has.

20. On a certain ground stand two poles 12 feet apart; the lesser pole is 35 feet in height and the greater, 40 feet. It is sought, if the greater pole will lean on the lesser, then in what part will it touch?

21. The following is a variant of one of Alcuin of York's problems (ca. 800) and the ancestor of the "jeep in the desert" problem: A certain gentleman

ordered that 90 measures of grain were to be transported from his house to another, 30 *leucas* distant. One camel was to carry the grain, carrying 30 measures on each journey. The camel eats one measure for each leuca. How can the grain be transported? How much grain will get to the second house?

22. Two men have a certain amount of money. The first says to the second, "If you give me 5 denarii, I will have 7 times what you have left." The second says to the first, "If you give me 7 denarii, I will have 5 times what you have left." How much money does each have?

23. A man went to a draper and bought a length of cloth 35 *braccia* long to make a suit of clothes. The draper told him that when the cloth was shrunk and clipped, every 7 braccia would shrink 1 braccia. The man took him at his word, but instead, the cloth shrank 1 for every 6 braccia. How much cloth did the man lack?

24. A man had four creditors. To the first he owed 624 *ducats*; to the second, 546; to the third, 492; and to the fourth, 368. It happened that the man defaulted and escaped, and the creditors found that his goods amounted to 830 ducats in all. In what ratio should they divide this, and what will be the share of each?

25. A mouse is at the top of a poplar tree 60 braccia high, and a cat is on the ground at its foot. The mouse descends ½ braccia a day, and at night it turns back ⅙ braccia. The cat climbs 1 braccia a day and goes back ¼ braccia each night. The tree grows ¼ braccia between the cat and the mouse each day, and it shrinks ⅛ braccia every night. In how many days will the cat reach the mouse, how much has the tree grown in the meantime, and how far does the cat climb?

26. I wish to find three numbers of such nature that the first and the second with one-half of the third makes 20, and the second and the third with one-third of the first comes to 20, and the third plus the first and one-fourth of the second also equals 20.

27. There were two men, of whom the first had three small loaves of bread and the other two. They walked to a spring, where they sat down and ate; and a soldier joined them and shared the meal, each of the three men eating the same amount. When all the bread was eaten, the soldier departed, leaving 5 *bezants* to pay for his meal. The first man accepted 3 of these bezants, since he had three loaves; the other took the remaining 2 bezants for his two loaves. Was this division fair?

28. A certain lion could eat a sheep in 4 hours; a leopard could do so in 5 hours; and a bear, in 6 hours. How many hours would it take for the three animals to devour a sheep if it were thrown in among them?

29. Suppose I tell you that I bought saffron in Siena for 18 lire a pound and took it to Venice, where I found that 10 ounces Siena weight are equivalent to 12 ounces in Venice, and 10 lire in Siena money are equal to 8 lire Venetian. I sell the saffron for 14 lire Venetian money a pound. I ask how much I gained in percent.

30. Suppose you have two kinds of wine. A measure of the poorer sort is worth 6 denarii. One of the better sort is worth 13 denarii. I wish to have a measure of wine worth 8 denarii. How much of each wine should I put in the mixture?

31. A certain man doing business in Lucca doubled his money there and spent 12 denarii. Thereupon, leaving, he went to Florence, where he also doubled his money and spent 12 denarii. Returning to Pisa, he doubled his money and spent 12 denarii, with nothing remaining. How much did he have in the beginning?

32. A certain man visited three fairs, carrying with him 10½ denarii. He doubled his money at each fair and also spent a certain amount—the same amount—at each fair. He returned home with no money. How much did he spend at the fairs?

33. A certain man invests 1 denarius at interest at such a rate that in 5 years he has 2 denarii, and in 5 years thereafter the money doubles. I ask you how many denarii he would gain from this 1 denarius in 100 years.

34. If 1000 pounds of pepper are worth 80 ducats, 16 grossi and ¼, what will 991 / ½ pounds be worth? [1 ducat = 24 grossi.]

35. Three men having denarii found a purse of 23 of denarii. The first man said to the second, "If I take this purse, I will have twice as much as you"; the second said to the third, "If I take this purse I will have three times as much as you"; and the third said to the first, "If I take this purse, I will have four times as much as you." How much did each one have?

36. Three men—Thomasso, Domenego, and Nicolo—went into a partnership. Thomasso put in 760 ducats on the first day of January 1472 and on the first day of April took out 200 ducats. Domenego put in 616 ducats on the first day of February 1472 and on the first day of June took out 96 ducats. Nicolo put in 892 ducats on the first day of February 1472 and on the first day

of March took out 252 ducats. On the first day of January 1475, they found that they had gained 3160 ducats, 13½ grossi. Required the share of each so that no one shall be cheated.

37. A merchant gave a university 2814 ducats on the understanding that he was to be paid back 618 ducats a year for nine years, at the end of which the 2814 ducats should be considered paid. What compound interest was he getting on his money?

38. A man buys a number of bales of wool in London; each bale weighed 200 pounds, English measure, and each bale cost him 24 *florins*. He sent the wool to Florence and paid carriage duties and other expenses amounting to 10 florins a bale. He wishes to sell the wool in Florence at such a price as to make 20% profit on his investment. How much should he charge a hundredweight if 100 pounds London are equivalent to 133 Florentine pounds?

39. Two men rent a pasture for 100 lires on the understanding that two cows are to be counted as being equivalent to three sheep. The first puts in 60 cows and 85 sheep; the second, 80 cows and 100 sheep. How much should each pay?

40. A merchant owes 500 pounds, to be paid in payments of 300 pounds in four months, 100 pounds at six months, and 100 pounds at 12 months. The debtor agrees to discharge the whole debt in one payment. Now the question is at what time the payment ought to be made, without damage to debtor or creditor. The creditor agrees to receive 6% per annum.

41. Given a triangle with one side measuring 20 units, the altitude drawn to that side is 6 units; the remaining sides are unknown, but one is twice the length of the other. What are the unknown sides of this triangle?

42. Given three mutually tangent circles of different radii, if the area of the triangle formed by connecting their centers is 367.424 square units and the radii of two circles are known to be 10 and 15 units, what is the radius of the third circle?

WHAT ARE THEY DOING?
Reckoning Monetary Exchange

One of the mathematical complexities faced by merchants in the Middle Ages and early Renaissance was the exchange of money. With the fall of the Roman Empire, the power and consistency of a central monetary agency collapsed. Without money, much trade was done by barter. Charlemagne, during his rule (768–814), introduced a monetary system based on a pound of silver as a standard unit of exchange. One pound was divided into 20 shillings or 240 pennies. The monetary unit, a pound, is still used by countries of the world today. The Arab empire developed a monetary system based on the use of gold, minting the *dirham* or *dinar*. Eventually, many trading cities in Italy developed their own mints and systems of currency. One of the most highly respected of such systems was that of the city of Venice, which functioned on the gold-backed *ducat*. However, the multitude of existing currencies of varying values and worth spawned commercial concern and very specific techniques for the mathematics of exchange.

The woodcut print from a fifteenth-century German arithmetic book shown in figure 10.1, illustrating a monetary exchange, prefaced a set of problems that dealt with the specifics of exchange. A typical example of such problems is the following:

> Seven pounds at Padua make 5 pounds at Venice, 10 pounds at Venice make 6 pounds at Nuremberg, and 100 pounds at Nuremburg make 73 pounds at Cologne; how many pounds at Cologne do 1000 pounds at Padua make?

The solution was attained by a process called *Regula del chatanina*, the "chain rule," which is carried out the aid of the following graphic scheme:

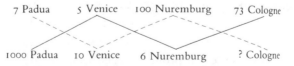

Thus, 1000 pounds at Padua are reduced to 312⁶/₇ pounds at Cologne.

Figure 10.1. Exchange of money as depicted in woodcut print from 1500 edition of Johannes Widmann's *Arithmetic*.

In the new growth of European trade, it was initially the Italian merchants who dominated the field and employed new reckoning techniques including the use of Hindu Arabic numerals to assist them in solving commercial problems. Specific techniques, such as the one above, were given Italian names. In addition, much of the mercantile vocabulary evolved from Italian terms and phrases. In Italian, a table on which computations were undertaken and money exchanged was a *banco*, from which the modern English word "bank" emerged. When a money changer was found to be corrupt, his table was physically broken—*bancoruptus*: he was bankrupt. Similarly, the word "endorse" comes from the Italian *endorso*, "to sign on the back." Bills of monetary exchange, which also originated at this time, had to be verified by a signature.

In the eighteenth century, when the United States of America came into being, similar problems with monetary exchange were encountered. Each of the 13 colonies had their own system of money, and the exchange of funds between colonies resulted in problems. The U.S. Congress established a federal monetary system in 1786. It was a decimal system, as explained in the excerpt shown in figure 10.2, from *The Scholar's Arithmetic; or, Federal Accountant* (Adams 1821, 80).

Aided by tables of exchange, students of the time were subjected to problems such as this one:

> Change 46 pounds 10 shillings 6½ pence entered in either currency to federal money.
>
> Answer: $155, 09 New England currency; $116, 317 New York currency; $124, 072 Pennsylvania currency.

Note that in the answer, the commas and the following spaces serve as modern decimal points. For example, the Pennsylvania currency would be read as 124 dollars, 7 cents, and 2 mils

Figure 10.2. Table showing decimal conversions for U.S. monetary system (1821).

The denominations are in a *decimal proportion*, as exhibited in the following

TABLE.

10 Mills		Cent,
10 Cents	make one	Dime,
10 Dimes		Dollar, *marked thus,* $
10 Dollars		Eagle.

The expression of any sum in Federal Money is simply the expression of a *mixed number* in decimal fractions. A dollar is the *Unit Money*; dollars therefore must occupy the place of units, the less denominations, as dimes, cents, and mills, are decimal parts of a dollar, and may be distinguished from dollars in the same way as any other decimals by a comma or separatrix. All the figures to the left hand of dollars, or beyond units place are eagles. Thus, 17 eagles, 5 dollars, 3 dimes, 4 cents, and 6 mills are written—

Eagles; or, Hundreds.
Doll's; or, Units. Tens.
Dimes; or, Tenth parts.
Cents; or, Hundredth parts.
Mills; or, Thousandth parts.

1 7 5,3 4 6

Of these, four are real coins, and one is imaginary.

The real coins are the Eagle, a gold coin; the Dollar and the Dime, silver coins; and the Cent, a copper coin. The Mill is only imaginary, there being no piece of money of that denomination.

There are half eagles, half dollars, double dimes, half dimes, and half cents, real coins.

A TABLE OF FOREIGN COINS, &c.—Showing their value in
the United States as established by acts of Congress—not
given in the preceding tables :—

SILVER.	D.	c.	m.
English or French Crown - - - -	1	10	0
Dollar of Spain, Sweden, and Denmark - -	1	00	0
Dollar of Mexico and S. A. S. - - -	1	00	0
Five-franc Piece - - - - - -	0	93	6
Franc - - - - - - - -	0	18	8
Pistareen - - - - - - -	0	20	0
Pound of Ireland - - - - - -	4	10	0
Pagoda of India - - - - - -	1	94	0
Tale of China - - - - - -	1	48	0
Millrea of Portugal - - - - - -	1	28	0
Ruble of Russia - - - - - -	0	66	0
Rupee of Bengal. - - - - - -	0	55	5
Guilder of the United Netherlands - - -	0	39	0
Mark-Banco of Hamburg - - - -	0	35	5
Livre of Francois . - - - - -	0	18	5
Gold Ducat of Russia - - - - -	2	00	0

Figure 10.3. Table for international monetary conversion from textbook of 1846.

Monetary conversion problems in nineteenth-century American arithmetic books were further complicated by the fact that the United States
was becoming an international commercial power. Conversion problems
involving trade also had to consider the value of foreign monies, particularly
those executed in gold and silver. Once again, tables of conversion were
supplied for the students' reference. The table of foreign silver conversion
shown in figure 10.3 is taken from the *Columbian Calculator* of 1845 (Ticknor 1845, 161).

11   Japanese Temple Problems

Illustration for the pouring of oil problem (problem 16 below), as given in *Jinkoki* (1643).

The mathematics books used in early Japan were imported Chinese classics such as the *Jiuzhang suanshu*. By the seventeenth century, however, special indigenous problems began to appear in Japan. These problems emerged during the Edo period (1603–1867) of Japanese history, when the country retreated into an imperially imposed state of isolationism. During this period of accompanying cultural introspection, the mathematician Yoshida Kōyū published a collection of 12 unsolved challenge problems. These problems were taken up and solved by readers who, in turn, posed their own challenge problems. Thus, a popular wave of problem solving and posing developed, based mainly on the solutions of complex geometric configurations and situations involving circles, ellipses and other common geometric curves. These problems, called *sangaku*, were solved by people from all social strata who in their pride of accomplishment posted their solutions and problems on inscribed wooden tablets and hung them in local Buddhist temples or Shinto shrines. The problems became known as Temple problems. These collections of mathematical expression stand as a testimony to the climate of mathematical creativity and problem solving ingenuity that existed in Japan at that time.

Problems

1. There is a circle from within which a square is cut; the remaining portion has an area of 47.6255 square units. If the diameter of the circle is seven more units than the square root of the side of the square, it is required to find the diameter of the circle and the side of the square.

2. A castle has n rooms in each of which there are seven samurai. Their total number, $7n$, leaves remainders of 9 and 15 when divided by 25 and 36, respectively. Find the least possible value for n.

3. On a day in spring a boy has gathered cherry blossoms under a cherry tree. Nearby, a poet is reading some of his poems aloud. As he reads, the boy counts out the cherry blossoms, one blossom for each word of a poem. After a number of poems (haiku) each 17 words in length, the boy has three blossoms remaining; after some 28-word poems, the boy has five blossoms left; after some 31-word poems, he has eight blossoms left. What is the least number of blossoms the boy has gathered?

4. The purchase price for an apple and an orange is 100 *yen*. When n oranges and $n+3$ apples are bought, the price is 520 yen. Find the number, n, of oranges and the price of one orange.

5. A boy gives 11 coins of equal denomination to a man, and the man finds that their total value in yen is 4 less than his age. The boy gives the man 9 coins of equal denomination, different from before, and the man finds that their total value in yen is 5 less than his age. What is the age of the man?

6. A set of four congruent circles whose centers form a square are inscribed in a right triangle ABC, where C is the right angle and serves as one corner of the square. Find their radii (r) in terms of the sides, a, b, and c, of the triangle.

7. The inscribed circle $O(r)$ of triangle ABC touches AB at D, BC at E, and AC at F. Find r in terms of AD, BE, and CF.

8. Circle $O(r)$ is inscribed in an isosceles trapezoid $ABCD$, where $AB = CD$ and $AD = BC$. Show that $4r = (AB)(CD)$.

9. A set of n disjoint, congruent circles $O^i(r)$ ($i = 1, 2, \ldots, n$) packs the surface of a sphere S so that each region of the surface exterior to the circles is bounded by arcs of three of the circles. Find the possible values of the number, n, of circles, and in each case express r in terms of the radius of S.

10. There is a regular pentagon. The length of its side is a. Find the area of the pentagon. Generalize your result for an nonagon.

11. Given, two circles: O_1 with radius r_1 and O_2 with radius r_2, where $r_1 > r_2$. They are tangent to each other and mutually tangent to a common line. If O_1 is tangent to the line at point A and O_2 is tangent at point B, show that $(AB)^2 = 4r_1r_2$.

12. In the figure at the left, if the radii of each inscribed circle is 1, what are the dimensions of the bounding rectangle?

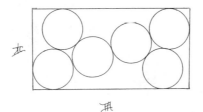

13. Given right triangle ABC, where C is the right angle, ellipse $O(a, b)$ is inscribed in it with its major axis parallel to BC. Calculate the semi-major axis, a, in terms of AC, BC, and b.

14. There is a log 18 feet long, the diameter of the extremities being 1 foot and 2.6 feet, respectively. This log is wound spirally with a string 75 feet long,

the coils being 2.5 feet apart. How many times does the string go around the log?

15. There is a mound of earth in the shape of a frustum of a cone. The circumferences of the bases are 40 measures and 120 measures, and the mound is 6 measures high. If 1200 cubic measures of earth are removed evenly from the top, what will be the remaining height?

16. A cooking oil peddler is selling oil. One evening on his way home, a customer asks him for 5 *sho* of oil. But the oil vendor has only 10 sho of oil left in his big tub and no way to measure the oil out except two empty ladles that can hold 3 and 7 sho. How does the oil peddler measure out 5 sho for the customer?

17. On January 1, a pair of mice appear in the house and bear 6 male mice and 6 female mice. At the end of January, there are 14 mice, 7 male and 7 female. On the first of February, each of the seven pairs has 6 male and 6 female mice, so that at the end of February, there are 98 mice, in 49 pairs. From then on, each pair of mice has six more pairs every month.

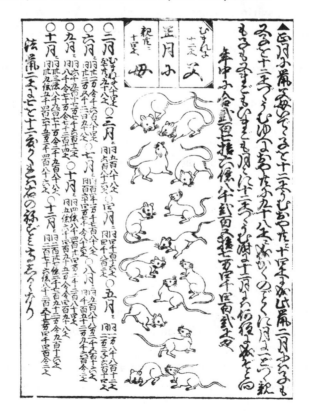

(a) Find the number of mice at the end of December.

(b) Assuming that the length of each mouse is 12 centimeters, if all the mice line up each biting the tale of the one in front, find the length of the mice train.

18. Given an equilateral triangle whose sides are 1 unit, as shown in the figure, draw three lines to the center, constructing three equal triangles, and in these triangles inscribe three circles. Show that the diameter of the circles is approximately 0.26794.

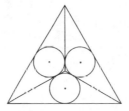

19. Given a pentagon whose sides are 1 unit, as shown in the figure, by drawing lines from the vertices to the center construct five triangles, and in each triangle inscribe a circle. Show that the diameter of the circles is approximately 0.50952.

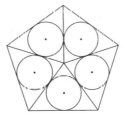

20. Let two circles be tangent to the same line and also to each other; let their radii be a and b, respectively; and let the points at which they touch the line be D and E. Show that $DE = 2\sqrt{ab}$.

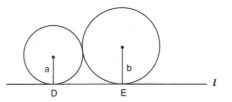

21. One night, some thieves stole a role of cloth from a shed. They are dividing cloth under a bridge when a passerby overhears their conversation: "If each of us gets 7 *tan*, then 8 tan are left over, but if each of us tries to take

8, then we are 7 tan short." How many thieves were there, and how long was the cloth?

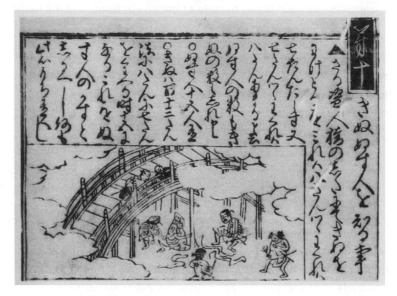

22. As shown in the figure, we divide an equilateral triangle *ABC* into three quadrilaterals that all have the same area. If the triangle has sides of length 14 and *G* is the center of the triangle, find the area of *ABC* and the length of the side of the small quadrilaterals.

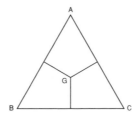

23. Civil servants A and B work in the town office. Civil servant A goes to work on every 12th day, and B goes to work on every 15th day. Today they meet each other in the office. How many days will pass before the next meeting?

24. A circle of radius *R* is inscribed in an isosceles triangle with sides equal to 12 and its base equal to 10. Find the length of *R*.

25. A melon stem grows 7 *sun* a day. A creeper stem grows 10 sun per day. In the same day, the melon stem grows down from a point on a cliff that is 90 sun high, and the creeper grows up from the bottom of the cliff. After how many days will the two stems meet?

26. Two circular roads A and B are tangent to each other at point P. Road A has a circumference of 48 kilometers and road B, 32 kilometers. A cow and a horse start walking from point P along roads A and B, respectively. The cow walks 8 kilometers per day and the horse walks 12 kilometers a day. How many days later do the cow and the horse meet again at point P?

27. Nine circles of radius r can be packed into a circle of radius R. Express r in terms of R.

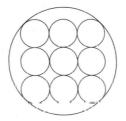

28. There is a field shaped like a doughnut. The outer circumference is 120 *ken*, while the inner circumference is 84 ken. A house sits in the middle of the field, so we cannot measure its diameter, but the distance between the two circumferences is 6 ken. Find the area of the field without using π.

29. Two circles of radius r are inscribed in a square and touch each other at the center. Each of two smaller circles with radius t touches the two sides of the square as well as the common tangent between the two larger circles. Find t in terms of r.

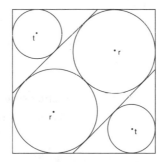

30. Here is 225.36 *koku* of rice. The government wants to distribute it to five classes of homes. Each second-class home gets 0.8 the amount of each of the 4 first-class homes. Each third-class home gets 0.8 times as much rice as each of the 8 second-class homes. Each of the fourth-class homes gets 0.8 times as much as each of the 15 third-class homes. Each of the 120 fifth-class homes receives 0.8 times the amount of the rice given to each of the 41 fourth-class homes. How much rice does each home and each class get?

111

31. A number of visitors, *N*, attend a shrine. We only know that

$\frac{7}{9}$ *N* is an integer and the last two digits are 68;

$\frac{5}{8}$ *N* is an integer and the last two digits are 60.

Find the least possible value for *N* that satisfies these conditions.

32. Two reeds of equal height project 3 *syaku* above the surface of a pond. If we draw the top of one reed 9 syaku in the direction of the shore so that its top is just touching the surface of the water, find the depth of the pond.

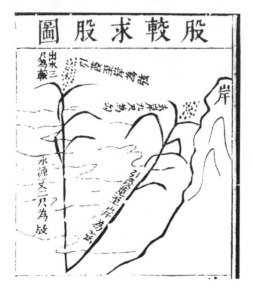

33. Just using the numbers 1, 2, and 3 three times each, make a magic square such that the sum of each row, each column, and each diagonal equals 6.

34. A circle of radius *r* inscribes three circles of radius *t*, the centers of which form an equilateral triangle of side 2*t*. Find *t* in terms of *r*.

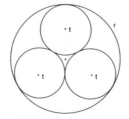

35. A circle of radius *r* is inscribed in an isosceles triangle whose base is 12 and whose sides are 10. Find the value of *r*.

36. A boy saves money in the following way: on the first day, he saves 1 *mon*, on the second day he saves 2 mon, and on the third day he saves $2^2 = 4$ mon. How much money will he have saved after 30 days?

37. A horse was stolen. The owner discovered the theft and began to chase the thief, who had already gone 37 *ri*. After the owner traveled 145 ri, he learned that the thief was still 23 ri ahead of him. After how many more ri will the owner catch up with the thief?

38. A number of identical balls are piled up in a trapezoid as shown below, where the base contains seven balls and the top contains the three balls. What is the total number of balls?

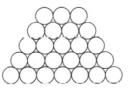

39. View the picture shown below and determine how many barrels are in the stack. Do not merely count the barrels. Use your knowledge of mathematics!

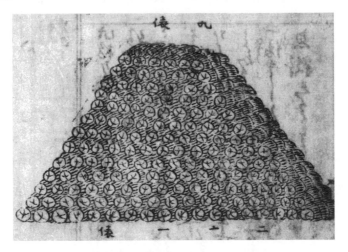

113

WHAT ARE THEY DOING?

Finding the Missing Numbers in Japanese Problems

Figure 11.1 depicts a problem in which one is asked to find the missing numbers from the "worm-eaten paper." Problems of this form, found in traditional Japanese mathematics, are called *mushikuisan*. Try some problems of this kind by choosing digits from the set of integers {0, 1, 2, . . . , 9} to make correct mathematical statements from the open exercises:

The area of a rectangle whose sides are 253 cm and [ ] [ ] cm is [ ] [ ] [ ] 39 cm².

[ ] [ ] 3 hits were scored in shooting 364 arrows in an archery contest, for an average score of 0.[ ] [ ].

The product 237 × [ ] [ ] = [ ] [ ] 47.

Figure 11.1. "Find the missing numbers," a traditional Japanese mathematics exercise.

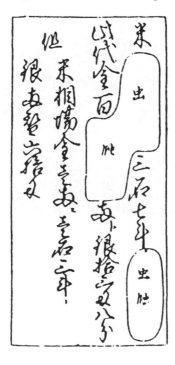

12   The *Ladies Diary*
(1704–1841)

Cover of the 1709 issue of the *Ladies Diary*, featuring a portrait of reigning Queen Anne. Her official patronage was sought but not given.

The *Ladies Diary* was a popular magazine published annually in England during the eighteenth and nineteenth centuries. Originally it was intended just for a female readership, but soon it became popular with the public in general. At the height of its popularity it sold 30,000 copies in a year. It was a sort of scientific digest for its times. One of the features of the *Diary* was a series of mathematical challenge problems to be solved by the readers. The answers and solutions for the problems would appear in the next issue. This magazine helped to promote mathematical problem solving among the middle class in England. Several collections of its mathematical problems and solutions were published separately. When the first U.S. mathematical journal, the *Mathematical Correspondent*, was founded in 1804, it was fashioned after the *Ladies Diary* and stressed problem solving. A selection of problems from the *Diary* are presented below.

Problems

1. Being employed to survey a field, which I was told was an exact geometrical square but by reason of a river running through it, I can only obtain partial measurements, I measure 9 yards from the west corner along the south side. Then sighting upon the northeast corner, I measure 18 yards along this line before turning and sighting back to the southeast corner, which I find at an angle of 28° 30′ from my path of previous sighting. From these measurements, find the area of the field.

2. A man hired a horse in London at 3 pence a mile. He rode 94 miles due west to Bristol, then due north to Chester, whence he returned toward London for 66 miles, which put him in Coventry. From Coventry he made a round trip to Bristol and then continued on to London. How much did he owe for the hire of the horse if the road from Coventry to Bristol meets the road from Chester to London at right angles?

3. Seven men held equal shares in a grinding stone 5 feet in diameter and agreed that each should use it until he had ground his share. What part of the wheel should each grind away?

4. Prove that if the sums of the squares opposite sides of any quadrilateral are equal, its diagonals intersect at right angles.

5. If you are h feet tall and walk all the way around the Earth, keeping to the same circumference, how much farther has your head gone than your feet when you complete the journey?

6. Being in a room opposite to the side of a window, the bottom of which was just the height of my eye, I observed that up to the edge of the window I could see 42 courses of bricks in a wall on the opposite side of the street; but moving towards the window 5 yards, I found that I could see 72 courses. Required the height of the window, supposing the breath of the street to be 12 yards, and 4 courses of brickwork to the foot in height.

7. Two circles of radii 25 feet intersect so that the distance between their centers is 30 feet. What is the length of the side of the largest square inscribable within their intersecting arcs?

8. There is a garden that is the shape of a rhombus whose side is 768.52 feet. Within the garden is an inscribed square flower bed whose side is 396 feet. What is the area of the garden?

9. A general who had served the king successfully in his wars, asked as a reward for his services, a *farthing* for every different file of 10 men each, which he could make with a body of 100 men. The king, thinking the request a very modest one, readily assented. Pray, what sum would it amount to?

10. You observe the edge of the cloud at an altitude of 20° and the sun above it at an altitude of 35°. The shadow cast by the edge of the cloud falls on an object that you know is 2300 yards from your position. How high is the cloud?

11. Given a semicircle, extend its diameter an arbitrary distance to point P, draw PC at a convenient angle intersecting the circle at points B and C. At B and C erect perpendiculars to PC and extend them to cut the diameter at D and E. Prove that if O is the circle's center, DO= OE.

12. A round pond sits in a rectangular garden. Its center is inaccessible; however, you know the distances from each corner to the center of the pond: 60, 52, 28, and 40 yards. What is the radius of the pond?

13. A gentleman has a garden of rectangular form and wants to construct a walk of equal width halfway around to take up half the garden. What must be the width of this walk? [Assume the garden has sides of lengths a and b.]

14. A lazy surveyor found the height of the tower without measuring any angle as follows. Some distance from the tower he sighted its top and angle he called α. Then he drew the complement of α on a piece of paper and walked forward until he sighted the tower at a right angle minus α. Next he drew an angle twice α and marched ahead on the same straight line until he saw the tower's top at 2α. If the distance between the first and second

stations was α and that between the second and third station b, show that y, the height of the tower, is given by $y^2=(b+3\alpha/2)/(b+\alpha/2)$.

15. A gentleman wishing to know how much it would cost him to fence a trapezoidal field at sixpence a pole and unable to manage the calculation himself, applied to the *Ladies Diary*. He gave the following information: the length of the larger base equals 1432 links; one base angle is 34° 17′, the other base angle is 54° 18′; the area of the field is 2.75 acres. How much will it cost to fence this field?

16. A circular vessel, whose top and bottom diameter are 70 and 92, and whose perpendicular depth is 60 inches, is so elevated on one side that the other becomes perpendicular to the horizon; required, what quantity of liquor, ale measure, will just cover the bottom when in that position.

17. A riddle:

> Odds bobs, ladies, what am I?
> I'm at a distance, yet am nigh;
> I'm high and low, round, short, and long,
> I'm very weak, and very strong.
> Sometimes gentle, sometimes raging.
> Now disgusting, now engaging.
> I'm sometimes ugly, sometimes handsome. . . .
> I'm very dirty, very clean,
> I'm very fat, and very lean,
> I'm very thick, and very thin,
> Can lift a stone, tho' not a pin. . . .

18. The following problem was proposed in 1838 and received many solutions, including one from Lewis Carroll in 1894: Upon the sides of triangle *ABC* the squares *ABDE*, *BCFG*, *ACHL* are constructed exterior to the triangle. Construct triangle *ABC* given *A′*, *B′*, *C′*, which are the intersection of *DE* and *HL*, *ED* and *FG*, *GF* and *LH*, respectively.

19. Given two circles that touch each other: one with center *O* and radius$=r_1$; the other center at *Q* and with radius$=r_2$. Let the common tangent to the circles be *PT*. Prove that $PT^2=4r_1r_2$.

20. A lady paid twice as much apiece for geese as she paid for ducks; and twice as much apiece for ducks as she paid for chickens, which cost together 1£ 13s 4 p. The sum of the squares of the number [of fowls] she bought was

326. What is the number of geese, ducks and chickens she bought? And what is the price of each?

21. Let *AC* be a semicircle with diameter *ABC*, where its radii equal *r*. Two other semicircles have centers *E* and *F* on *AC*. A circle centered at point *D* is tangent to these three semicircles. Given that *AE* = *BF* = *r*/2, show that the radius of the circle centered at *D* is *AB*/3.

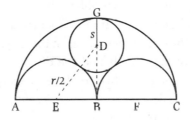

22. Another riddle (remember that this was written in the eighteenth century):

> My Parent brought me forth without a Head
> Then lay I useless motion less and dead
> But some time after moll ingeniously
> By's godlike Art he plac d ten Heads on roe
> I taking Huff at cruel Blows set out
> And boldly range the Country round about
> To Cities Towns and Villages I roam
> And well attended am where e er I come
> Why sliou d I not I much deserve their care
> Tho carried yet a mighty Weight I bear
> When thro the Streets I pass in darkest Nights
> I make young Sparks attend me with their Lights
> But such a Shape as mine I m sure was never
> I march along with Head and Heel together
> And am so low of Stature so minute
> I can t avoid being trampled under Foot.

WHAT ARE THEY DOING?
Mathematical Riddles

Throughout history, mathematics problems have frequently been posed in poetic form or as riddles. In the Rhind Mathematical Papyrus (1650 BCE) we find the riddle of the cats in sacks; later in the Greek Anthology, riddles concerning the age of individuals and the distribution of apples appear. When Alcuin of York composed his *Problems to Sharpen the Young* (800), many of the problems he devised were given in riddle form. Fibonacci also used riddles in his *Liber Abaci* (1202).

Perhaps the concept of a riddle added another attraction to problem solving; first, the riddle had to be solved as to the mathematical problem it contained, and then the mathematical problem itself had to be solved. Another reason for using riddles was that they were more genteel in challenging the reader. They could serve as an object of group discussion and a concern for a broader audience, avoiding intellectual confrontation with a single person. A use of riddles heightened the fact that mathematical problems were more than just exercises but were also intellectual activities. In many of the books and periodicals of the eighteenth and nineteenth centuries, problems were posed in riddle form. Below are two such riddles. Can you solve them without looking at the answers?

> One ev'ning I chanc'd with a tinker to sit,
> Whose tongue ran a great deal too fast for his wit:
> He talked of his art with abundance of mettle;
> So I asked him to make me a flat bottom kettle.
> Let the top and the bottom diameters be,
> In just such proportion as five is to three:
> Twelve inches the depth I propos'd, and no more;
> And to hold in ale gallons seven less than a score.
> He promised to do it, and straight to work went;
> But when he had done it he found it too scant.
> He altered it then, but too big he had made it;

For though it held right, the diameters fail'd it;
Thus making if often too big and too little,
The tinker at last had quite spoiled his kettle;
But declares he will bring his said promise to pass,
Or else that he'll spoil every ounce of his brass:
Now to keep him from ruin, I pray you find him out
The diameter's length, for he'll never do't I doubt. (*Ladies Diary*,
  1711)

Answer: bottom, 14.44401 inches; top, 24.4002 inches.

A gentleman whilst walking in his ground,
A stone, of shape and size uncommon found;
Which having with a curious pleasure viewed
A strong desire to measure it ensued:
This by the judgment of an artist done,
Its form was found to be an upright cone;
The slaunting side of which was eighteen feet,
The base diameter fourteen compleat.
He strait have orders that should be sent
Into his garden for an ornament;
Willing to have a room made in the same,
In one of these three forms which he should name:
A cube, a cylinder, or a hemisphere,
The greatest possible the stone can bear.
A skillful artist has proposed to cut
The same, at 20 pence the solid foot
The gentleman, all needless charge to save,
Some mathematical advice would have.
By which he may with satisfaction see,
Which has the least content of all the three;
That he may fix the form accordingly:
He likewise doth desire a just account
To what it will in sterling coin amount. (*Gentleman's Magazine*,
  1741)

Answer: a cube; cost, £19 18s 2p.

13 Nineteenth-Century Victorian Problems

A photo of well-known Victorian writer Lewis Carroll, author of *Alice's Adventures in Wonderland*, taken in 1863. His real name was Charles Lutwidge Dodgson, and he was a mathematician by profession. Dodgson wrote on logic and incorporated logic puzzles into his Alice stories.

Nineteenth-century Great Britain experienced a period of dramatic mathematical innovations. Early in the century, a group of students at Cambridge University founded the Analytical Society. This society advocated for the adoption of a Leibnizian calculus replacing the more awkward methods of fluxions as introduced by Isaac Newton. This reform brought the British scientific and mathematical community closer to their Continental peers in the areas of mathematical analysis. As if from a Dickensian playbill, a cast of intriguing and colorful mathematical characters emerged on the scene.

George Peacock (1791–1858), an original member of the Analytical Society and an active social reformer, formalized algebra, earning the appellation "the Euclid of Algebra." Indeed, most of Victorian work in mathematics would focus on developments in algebra and logic. George Boole (1815–64), a self-educated son of a cobbler, formalized the study of logic and devised an algebra of sets. He was related by marriage to the explorer Sir George Everest, for whom Mount Everest was named. Boole's mathematical efforts were supplemented by the work of Augustus De Morgan (1806–71) and John Venn (1834–1923). Charles Ludwidge Dodgson wrote on logic and, under the pseudonym Lewis Carroll, composed stories for children. This author of *Alice in Wonderland* was also an amateur photographer noted for his photos of scantily clad young girls. The Irish astronomer royal and child linguistic prodigy William Rowan Hamilton (1805–65), in his work with quaternions, devised a noncommutative algebra. Charles Babbage (1791–1871), "the father of modern computing," constructed gear-driven analytic engines to perform computations for the British navy. While sound in theory and design, these computers lacked the necessary mechanical precision to achieve their desired tasks, and the navy withdrew its support for the project. Babbage was assisted in his work by the lovely young courtesan and daughter of the poet Lord Byron, Ada Lovelace (1815–52). Countess Lovelace has been credited as being the first computer programmer and was dubbed the "enchantress of numbers" by her mentor.

These Victorian contributions to mathematics resulted in new outlooks for the discipline, stimulated new explorations, and resulted in exciting advancements that would have far-reaching effects into the twentieth century.

Problems

1. Wanting to know the breadth of a river, I measured a base of 500 yards in a straight line close by one side of it; and at each end of this line, I found angles subtended by the other end and a tree close to the bank on the other side of the river to be 53° and 79°12'. What is the perpendicular breadth of this river?

2. What number which being increased by ½, ⅓, and ¼ of itself, the sum shall be 75?

3. What is the perpendicular height of a cloud when its angles of elevation were 35° and 64° as taken by two observers at the same time, both on the same side of it, and in the same vertical plane; the distance between them is 880 yards. And what is the cloud's distance from these two observers?

4. How high above the earth must a person be raised, that he [or she] may see one-third of its surface?

5. Find the area of the elliptic segment cut off parallel to the shorter axis when the height of the segment is 10 and the axes of the ellipse are 25 and 35.

6. After a terrible battle it is found that 70% of the soldiers have lost an eye, 75% an ear, 80% an arm, and 85% a leg. What percentage of the combatants must have lost all four?

7. Two travelers, starting at the same time from the same point, travel in opposite directions around a circular railway. Trains start each way every 15 minutes, the easterly ones going around in 3 hours, the westerly in 2. How many trains did each traveler meet on his circuit, not counting trains met at the terminus itself?

8. An oblong garden is a half yard longer that it is wide and consists entirely of a gravel walk, spirally arranged, 1 yard wide and 3630 yards long. Find the dimensions of the garden.

9. Two circles of radii 25 feet intersect so that the distance between their centers is 30 feet. What is the length of the side of the largest square inscribable within their intersecting arcs?

10. A round pond sits in a rectangular garden. Its center is inaccessible; however, you know the distances from each corner to the center of the pond: 60, 52, 28, and 40 yards. What is the radius of the pond?

11. A man 5 feet high, standing at the base of a square pyramid, sees the sun rise over one of the edges of the square, halfway along it. Show that if a and b are the distances from the two nearest corners, and θ is the altitude of the sun, the height of the pyramid is $10 + \tan \theta \sqrt{\frac{1}{2}(5a^2 - 2ab + b)}$ feet.

12. A cliff with a tower on its edge is observed from a boat at sea; the elevation of the top of the tower is 30°. After rowing towards the shore a distance of 500 yards in the plane of the first observation, the elevations of the top and bottom of the tower are 60° and 45°, respectively. Find the height of the cliff and the tower.

13. Given two circles tangent at the point P with parallel diameters AB, CD, prove that APD and BPC are straight lines.

14. Three vertical posts are placed at an interval of one mile along a straight canal, each rising to the same height above the surface of the water. The line of vision joining the tops of the two extreme posts cuts the middle post at a point 8 inches below the top. Find to the nearest mile the radius of the earth.

15. One man and two boys can do in 12 days a task which could be finished in 6 days by three men and a boy. How long would it take a single man to do it?

16. Landowner A distributes 180 coins in equal sums among a certain number of workers. Landowner B distributes the same sum, but gives each of his workers 6 coins more that A and has fewer workers than does A. How much does A give to each worker?

17. Two men start walking at the same time and travel a distance of 10 miles. One goes two miles per hour faster than the other and completes the journey five-sixths of an hour sooner. How fast did each man travel?

18. The sum of the ages of a father and son is 100 years. Also, one-tenth of the product of their ages, in years, exceeds the father's age by 180. How old are they?

19. A number composed of three digits in base 7 has the same three digits in reverse order when expressed in base 9. What is the number?

20. Two post boys, A and B, 59 miles distance from one another, set out in the morning in order to meet. A rides 7 miles in two hours, and B, 8 miles in three hours; and B begins his journey one hour later than A. Find what number of miles A will ride before he meets B.

21. Having been given the perimeter and perpendicular of a right angle, it is required to find the triangle.

22. A certain merchant increases the value of his estate yearly by one-third; he also spends 100 pounds yearly on his family; after three years he finds the value of his estate has doubled. What was he originally worth?

23. If 20 men, 40 women, and 50 children receive £350 for seven weeks of work, and two men receive as much as three women or five children, what does a woman receive for a week's work?

24. The number of disposable seamen at Portsmouth is 800; at Plymouth, 756; and at Sheerness, 404. A ship is commissioned; its complement is 490 seamen. How many must be directed from each place so as to take an equal proportion?

25. If five pumps each having a length of stroke of 3 feet, working 15 hours a day for 5 days, empty water out of the mine, how many pumps with length of stroke 2½ feet, working 10 hours a day for 12 days, will be required to empty the same mine if the strokes of the former set of pumps are performed four times as fast as the other?

26. If 12 horses can plow 96 acres in six days, how many horses could plow 64 acres in eight days?

27. If 44 cannons firing 30 rounds per hour for three hours a day consumed 300 barrels of powder in five days, how long will 400 barrels last for 66 cannons firing 40 rounds per hour for five hours a day?

28. A starts a business with the capital of £2400 on 19 March, and on 17 July he admits a partner, B, with a capital of £1800. Their profits amount to £943 by 31 December. What is each person's share?

29. Given a triangle with one side of 20, the altitude drawn to that side is 6; the remaining sides are unknown, but one is twice the length of the other. What are the unknown sides of this triangle?

30. Given three mutually tangent circles of different radii, if the area of the triangle formed by connecting their centers is 367.424 units and the radii of two circles are known to be 10 and 15 units, what is the radius of the third circle?

WHAT ARE THEY DOING?

Illustrations from Early Problems

Mathematics problems were some of the first written and/or printed content to be accompanied by illustrations. In many cases, the illustrations are merely descriptive—that is, they reinforce the situation described rather than amplify the mathematics under consideration. They also possess a visual impact and attract a reader's attention. The illustration from Robert Recorde's *Pathway to Knowledge* (1551), shown in figure 13.1, does both.

Figures 13.2–13.3 are illustrations of two popular historical problems: the "broken bamboo" problem and the "river crossing" problem. The broken bamboo problem as stated in Milne's *Standard Arithmetic* (1892) is as follows:

A tree, broken off 21 feet from the ground, and resting on the stump, touches the ground 28 feet from the base of the stump. What was the height of the tree? (p. 333)

Several much older versions of this problem are shown in figure 13.2: a twelfth-century Indian version, a Chinese version of 1263, and an Italian version of 1491.

Figure 13.1. The scaling of a fortification was a popular visual theme for applications of the Pythagorean theorem.

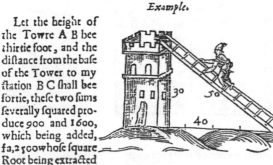

Example.

Let the height of the Towre A B bee thirtie foot, and the diſtance from the baſe of the Tower to my ſtation B C ſhall bee fortie, theſe two ſums ſeverally ſquared produce 900 and 1600, which being added, fa,2 500 whoſe ſquare Root being extracted is fiftie, the length of the Diagonall or ſcaling Ladder for that place which is the ſide A C.

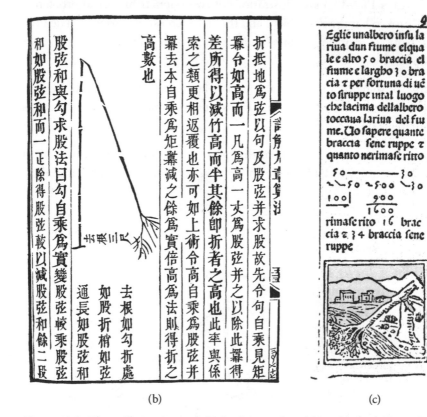

Figure 13.2. Three illustrations of the broken reed problem: (a) from the Indian classic, Bhaskara's *Lilavati* (1150); (b) from the Chinese *Jiuzhang suanshu* (ca. 100); and (c) from Filippo Calandri's *Arithmetic* (1491).

Figure 13.3. The boatman takes the goat across the water while the wolf and sheaf of wheat wait their turn.

The river crossing problem—in this case, involving a wolf, a goat, and a sheaf of wheat—is illustrated in a thirteenth-century manuscript reproduced in figure 13.3.

14   Eighteenth- and Nineteenth-Century American Problems

A nineteenth-century calendar engraving depicting an American agricultural scene, fall harvest.

Because America was founded as a British colony, its schools followed British practices and used arithmetic books that were popular in England at that time. However, after independence in 1776, American authors wrote arithmetic books designed for the needs of their emerging country. Many of the problems in these books are classical—that is, they can be found in most books of that period—yet many other problems reflect on life in this new country where agriculture was the main occupation and a merchant class was appearing. As the United States of America moved into the twentieth century, it became an industrial power, and later mathematical problems would reflect the needs of a different society.

Problems

1. Two merchants, A and B, loaded a ship with 500 hogsheads (hhds) of rum; merchant A loaded 350 hhds, and B the rest. In a storm the seamen were obliged to throw overboard 100 hhds. How much must each sustain of the loss?

2. If a ball 6 inches in diameter weighs 32 pounds, what will be the weight of a ball 3 inches in diameter?

3. Suppose a lighthouse is built on the top of a rock. The distance between a place of observation and that part of the rock level with the eye is 620 yards; the distance from the top of the rock to the place of observation is 840 yards; and from the top of the lighthouse, 900 yards. Find the height of the lighthouse.

4. A bridge is built across a river in 6 months by 45 men. It is washed away by the current. Required, the number of workmen sufficient to build another of twice as much worth in 4 months.

5. Gunmetal of a certain grade is composed of 16% tin and 84% copper. How much tin must be added to 410 pounds of this gunmetal to make a composition of 18% tin?

6. Knowing the base, b, and the altitude, a, of a triangle, find the expression for a side of the inscribed square.

7. Having been given the lengths, a and b, of two straight lines drawn from the acute angles of a right triangle to the middle of the opposite sides, determine the length of those sides.

8. A farmer sold a team of horses for $440 but did not receive his pay for them until 1 year 8 months after the sale. He had at the same time another

offer of $410 for them. Did he gain or lose by the sale and by how much, money being worth 6% per year?

9. A lady has two silver cups and only one cover for both. The first cup weighs 16 ounces, and when it is covered, it weighs three times as much as the second cup; but when the second cup is covered, it weighs four times as much as the first. What is the weight of the second cup and cover?

10. A speculator bought stock at 25% below par and sold it at 20% above par. He gained $1560. How much did he invest?

11. A teacher agreed to teach nine months for $562.50 and his board. At the end of the term, on account of two months' absence caused by sickness, he received only $409.50. What was his board worth per month?

12. There are four companies, in one of which there are 6 men, in another 8, and in each of the remaining two, 9 men. How many ways can a committee of 4 men be composed by choosing one man from each company?

13. What is the sum of the following series, carried to infinity: 11, $11/7$, $11/49$, . . . ?

14. If an equilateral triangle whose area is equal to 10,000 square feet is surrounded by a walk of uniform width, and the walk is equal to the area of the inscribed circle, what is the width of the walk?

15. A square circumscribed about a given circle is double in area to a square inscribed in the same circle. True or false? Prove your answer.

16. What proportions of sugar at 8 cents, 10 cents, and 14 cents per pound will compose a mixture worth 12 cents per pound?

17. A fellow said that when he counted his nuts by twos, threes, fours, fives, and sixes, there was still one left over; but when he counted them by sevens, they came out even. How many did he have?

18. Given, a right triangle where you know the length of the base and the sum of the perpendicular side and the hypotenuse. Find an expression for the lengths of the perpendicular side and the hypotenuse.

19. If $80 worth of provisions will serve 20 men for 25 days, what number of men will the same amount of provisions serve for 10 days?

20. A ladder is placed perpendicular to the plane of the horizon, and in coincidence with the plane of an upright wall. If the base of the ladder be drawn along the horizontal plane, in a direction perpendicular to the plane of the wall, with the top of the ladder sliding downwards against the wall, it is required to find the equation of the curve which is the locus of a point taken anywhere on the ladder.

21. The highest point of the Andes is about 4 miles above sea level. If a straight line from this point reaches the surface of the water [sea] at a distance of 178.25 miles, what is the diameter of the Earth?

22. Suppose a ladder 60 feet long is placed in a street so as to reach a window 37 feet above the ground on one side of the street and, without moving it at the foot, will reach a window 23 feet high on the other side of the street. How wide is the street?

23. A number is required that the square shall be equal to twice the cube.

24. A man filled his store with several sorts of grain, of which ½ was wheat, ¼ rye, ⅛ Indian corn, and ⅒ oats; and he had 8 barrels of barley. How much of each did he have and what was the total?

25. Suppose the area of an equilateral triangle be 600. The sides are required.

26. Find the diameter of a container in the form of a cylinder which is 60 inches high and that will hold 100 more units of wine than a square container of the same height, the sum of whose sides is equal to the circumference of the cylinder. [1 wine gallon = 231 in³.]

27. Four men owned £90 between them, so that if to the first man's money you add Z amount, it equals the second man's money diminished by Z, and the third man's money multiplied by Z, and the fourth man's divided by Z. What was each man's part of the £90?

28. There are two numbers which are to each other as 5 and 6, and the sum of their squares is 2196. What are the numbers?

29. A water tub holds 73 gallons; the pipe which fills it usually admits 7 gallons in 5 minutes; and the tap discharges 20 gallons in 17 minutes. Now supposing these both be carelessly left open, and the water to be turned on at 4 o'clock in the morning; a servant at 6, finding the water running, closes the tap. In what time after this accident will the tub be full?

30. Three equal circumferences with radii of 6 inches are tangent to each other. Compute the area enclosed between them.

31. A circle, a square, and an equilateral triangle all have the same perimeter, equal to 1 meter. Compare their areas.

32. If an arc of 45° on one circumference is equal to an arc of 60° on another circle, what is the ratio of the areas of the circles?

33. The steamer *Katie* leaves the wharf at New Orleans and runs an average speed of 15 mph. When the *Katie* has gone 25 miles, the steamer

R. E. Lee leaves the wharf and runs the average speed of 18 mph. How far will the *Lee* go before she overtakes the *Katie*?

34. If 40 oranges are worth 60 apples, and 75 apples are worth 7 dozen peaches, and 100 peaches are worth 1 box of grapes, and 3 boxes of grapes are worth 40 pounds of pecans, how many peaches can be bought for 100 oranges?

35. Two persons sat down to play for a certain sum of money and agreed that the first who got three games would be the winner. One of them won two games and the other, one game; but being unwilling to continue, they resolved to divide the stakes. How much should each person receive?

36. A man plants 4 kernels of corn, which at harvest produce 32 kernels: these he plants the second year. Now supposing the annual increase to continue eightfold, what would be the produce of the 15th year, allowing 1000 kernels to a pint? (Express the answer in bushels.)

37. A man agreed to pay for 13 valuable houses worth $5000 each. What would the last amount to, reckoning 7 cents for the first, four times 7 cents for the second, and so on, increasing the price four times on each to the last. Did he gain or lose by the bargain and how much?

38. A gentleman has bought a rectangular piece of land whose perimeter is to be 100 rods; and he is to pay $1 for each rod in the length of the land and $3 for each rod in the breath of the land. It is required to determine the length and breadth so that the quantity of land may be had at the cheapest price possible.

39. Two officers each have a company of men; the one has 40 fewer than the other. They divide among their men 1200 crowns. How many men are there in each company if the officer who had less gave 5 crowns more to each of his men than the officer who had more?

40. Determine by using algebra the number of degrees in the angle θ where $\cos \theta = \tan \theta$.

41. If $X^X = 100$, what is the value of X?

42. Suppose General [George] Washington had 800 men and was supplied with provisions to last 2 months but he needed to feed his army for 7 months. How many men must leave so that the remaining soldiers will be fed?

43. The three sides of a triangular piece of land, taken in order, measure 15, 10, and 13 chains, respectively; it is required to divide it into two equal parts by a line constructed parallel to the second side. What will be the length

of the division line and its distance from the place of beginning, measured on the first side?

44. A man, his wife, and two sons desire to cross a river. They have a boat that will carry only 100 pounds. The man weighs 100 pounds, the wife 100 pounds and the sons each 50 pounds. How can they all cross this river in the boat?

45. A lady being asked how old she was at the time of her marriage, replied that the age of her oldest son was 13; that he was born 2 years after her marriage; that when she married the age of her husband was three times her own; and that now her husband was twice as old as herself. How old were she and her husband when they married?

46. The perimeters of two similar triangles are 45 and 135 inches, respectively. One side of the first triangle has length of 11 inches and a second side, 19 inches. Find the lengths of the sides of the second triangle.

47. At what time between 4 and 5 o'clock will the hands of the clock be together?

48. The sum of the two digits of a two-digit number is 9. If 45 is subtracted from the number, the result will be expressed by the digits in reverse order. Find the number.

49. Two bicyclists travel in opposite directions around a quarter-mile track and meet every 22 seconds. When they travel in the same direction on this track, the faster passes the slower once every 3 minutes and 40 seconds. Find the rate of each rider.

50. Two cogwheels, one having 26 cogs and the other 20 cogs, run together. In how many revolutions of the larger wheel will the smaller gain in 12 revolutions?

51. A man bought a number of sheep for $225; 10 of them having died, he sold four-fifths of the remainder for cost and received $150 for them. How many did he buy?

52. I owe a man the following notes: one of $800 due May 16; one of $660 due on July 1; one of $940 due September 29. He wishes to exchange them for two notes of $1200 each and wants one to fall due June 1. When should the other be due?

53. A father willed his estate, valued at $40,000, to his three children in proportion as follows: John ⅓, Henry ¼, and Katie ⅕. Before the settlement was made, Henry died. What should John and Katie each receive?

54. X, Y, and Z hired a pasture for the season for $90. X pastured 9 head of mules for 150 days, Y pastured 11 head for 110 days, and Z pastured 24 head of mules for 160 days. How much is each to pay?

55. A man invested $100 in 100 head of livestock, consisting of calves, goats, and pigs. The price of each is as follows: calves, $10 apiece; goats, $1 apiece; and pigs, $0.12½ apiece. How many of each did he buy? [Note: The answer to this problem is obtained using an optimum strategy; the farmer is getting the "best deal" possible. Can you figure out the solution strategy?]

56. Three persons bought a sugar loaf in the form of a perfect cone 25 inches high and agreed to divide it equally by sections cut parallel to the base. What was the slant height of each one's share, the base being 12 inches in diameter?

57. A California miner has a spherical ball of gold 2 inches in diameter, which he wants to exchange for spherical balls 1 inch in diameter. How many of the smaller spheres should he receive?

58. A general formed his men into a square—that is, an equal number in rank and file—and he found that he had an excess of 59 men. Then, he increased both rank and file by one man equally and, forming a square, found that he was 84 men short. How many men did he have at his command?

59. What will the diameter of a sphere be when its volume and surface area are expressed by the same number?

60. What must be the inside dimensions of a cubical box to hold 200 balls, each 2½ inches in diameter?

61. Required, the side of the largest equilateral triangular beam that can be hewn from a piece of round timber 36 inches in diameter.

62. If two men or three boys can plow an acre in one-sixth of the day, how long will it require three men and two boys to plow it?

63. If out of a cargo of 600 slaves, 200 die during the passage of six weeks from Africa to the West Indies, how long must the passage be that one-half the cargo may perish? Suppose the degree of mortality to be the same throughout the passage, that is, the number of deaths at any time to be proportional to the living at the same time.

64. A drover sold cows and sheep for $6105. If he received 65% more for the cows than for the sheep, how much did he get for the cows?

65. A, B, and C bought a lot of merchandise on speculation for $3000. A paid $1500, B paid $900, and C paid $600. It was agreed that each one's interest in the goods should be proportional to the amount of money invested.

They then sold to D one-fourth interest in the goods for $1800; and A, B, and C agreed to adjust the matter so that each of them should be one-third owner in the remainder of the goods. What was the financial arrangement between A, B, and C?

66. Three young gentleman, A, O, and P, contracted for the use of a carriage for 30 days for $100, which they agreed to pay in proportion to the number of days that each had used it. A used it for 9 days, O used it for 14 days, and P for 6 days. For 1 day it was not in use. What is the amount owed by each?

67. Two newsboys, James and Henry, on taking their seats to eat some cakes they had purchased, were requested by another member of their fraternity named Dick to allow him to dine with them. To this request, they cheerfully assented, and the three ate cakes. When they had finished, Dick laid on the table $0.40 to pay for his share of the dinner. How much of this $0.40 was James, who contributed five cakes to the dinner, to receive; and how much should Henry, who contributed three cakes, receive?

68. A young hoodlum, a modern evolvement of the human race, stole a basket of peaches and divided them among three brothers, also hoodlums, and himself as follows: to the first, he gave one-quarter of the whole number and one-quarter of a peach more; to the second, he gave one-third of what remained and one-third of a peach more; to the third, he gave one-half of what remained and one-half of a peach more. The thief retained what was left for himself, which was one-half the number he gave to the first hoodlum. What was the number of peaches stolen, and how many did each receive?

69. A wine merchant has wine worth $1.10 per gallon, $1.80 per gallon, and $2.50 per gallon. He wishes to mix this wine with water so that the compound mixture will be worth $1.50 per gallon. What will be the proportional quantities of wine and water?

70. A tree 100 feet tall is broken in the storm, and the top touches the ground 40 feet from the foot of the tree. What is the length of the portion broken off?

71. A corncrib filled with corn on the ear is 20 feet×8 feet at the top, 16 feet×6 feet at the bottom, and 11 feet high. Allowing 2 even bushels of corn on the ear to make 1 bushel of shelled corn, how many bushels of shelled corn are there in the crib?

72. A planter, being asked how many head of mules he had, replied that if he had as many more, one-half as many more, and seven mules and a half, he would have 100. How many did he have?

73. Three footmen, A, B, and C, start together and travel in the same direction around an island that is 72 miles in circumference. Footman A travels 6 miles a day, B travels 8 miles a day, and C travels 12 miles a day. How long before they will all be together again?

74. A livestock dealer sold two horses, each at the same price. On one horse he gained 25%, and on the other he lost 25%. His net loss on the sales was $12. What was the cost of each horse?

~~~

WHAT ARE THEY DOING?
Nineteenth-Century Copybook Assignments

The page shown in figure 14.1 contains several examples on the topic of arithmetic progressions. In early America, textbooks were scarce and, if available, quite expensive; in many cases they remained only in the possession of the teacher. In such instances, the teacher would dictate lessons to the students, and they would copy them in a copybook. If this copybook concerned mathematics, it was often called a "cipher book." In such books, students would strive to demonstrate a high level of achievement in the subject being considered and to provide examples of their excellent penmanship. (Remember, this is still the time of the quill or steel-tipped pen.) Cipher books often also served as personal diaries or records of family financial accounts. Examining the contents of such books provides an insight into both the mathematics of the times and the life of the people involved. The reader might try to solve the indicated problems.

One of the most historically interesting and important cipher book pages in existence is shown in figure 14.2, in which the student was practicing his multiplication knowledge. First, can you determine the historical significance of this page? Next, can you figure out what the crosses with numbers in them represent?

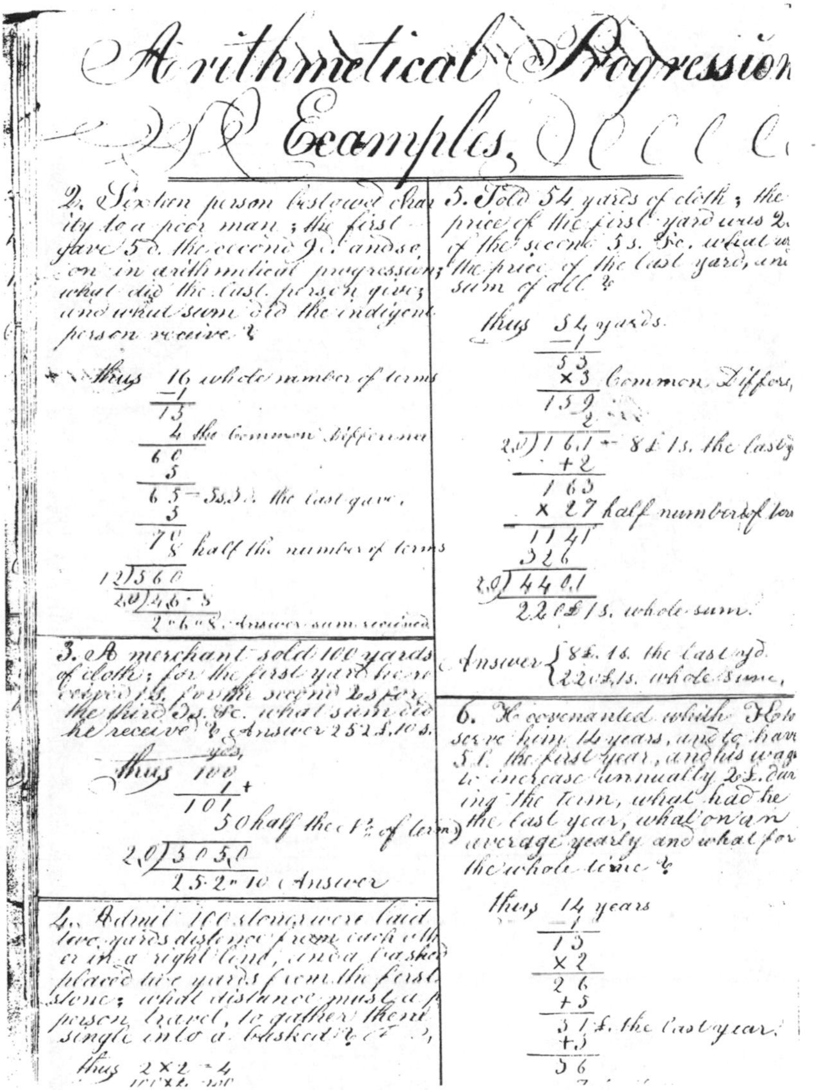
Arithmetical Progressions
Examples.

2. Sixteen person bestowed charity to a poor man; the first gave 5d. the second 9d. and so on in arithmetical progression; what did the last person give; and what sum did the indigent person receive?
thus 16 whole number of terms
-1
15
4 the common difference
60
5
65=5s.5d. the last gave.
5
70
8 half the number of terms
12)560
20)46-8
2-6-8 Answer sum received

3. A merchant sold 100 yards of cloth; for the first yard he received 1s. for the second 2s. for the third 3s. &c. what sum did he receive? Answer 252£.10s.
thus 100 yds.
1+
101
50 half the No. of terms
2,0)505,0
252-10 Answer

4. Admit 100 stones were laid two yards distance from each other in a right line, and a basket placed two yards from the first stone; what distance must a person travel, to gather them single into a basket?
thus 2×2=4

5. Sold 54 yards of cloth; the price of the first yard was 2s. of the second 5s. &c. what was the price of the last yard, and sum of all?
thus 54 yards.
-1
53
×3 Common Differ.
159
2
2,0)16,1=8£ 1s. the last y.
+2
163
×27 half number of terms
1141
326
2,0)440,1
220£ 1s. whole sum.
Answer {8£. 1s. the last yd. / 220£.1s. whole sum.

6. He covenanted with H to serve him 14 years, and to have 5£. the first year, and his wages to increase annually 2£. during the term, what had he the last year, what on an average yearly and what for the whole time?
thus 14 years
-1
13
×2
26
+5
31£. the last year.
+5
36

Figure 14.1. "Arithmetical Progressions," a page from the cipher book of Daniel Danner, Manheim County, Lancaster, Pennsylvania, January 29, 1819.

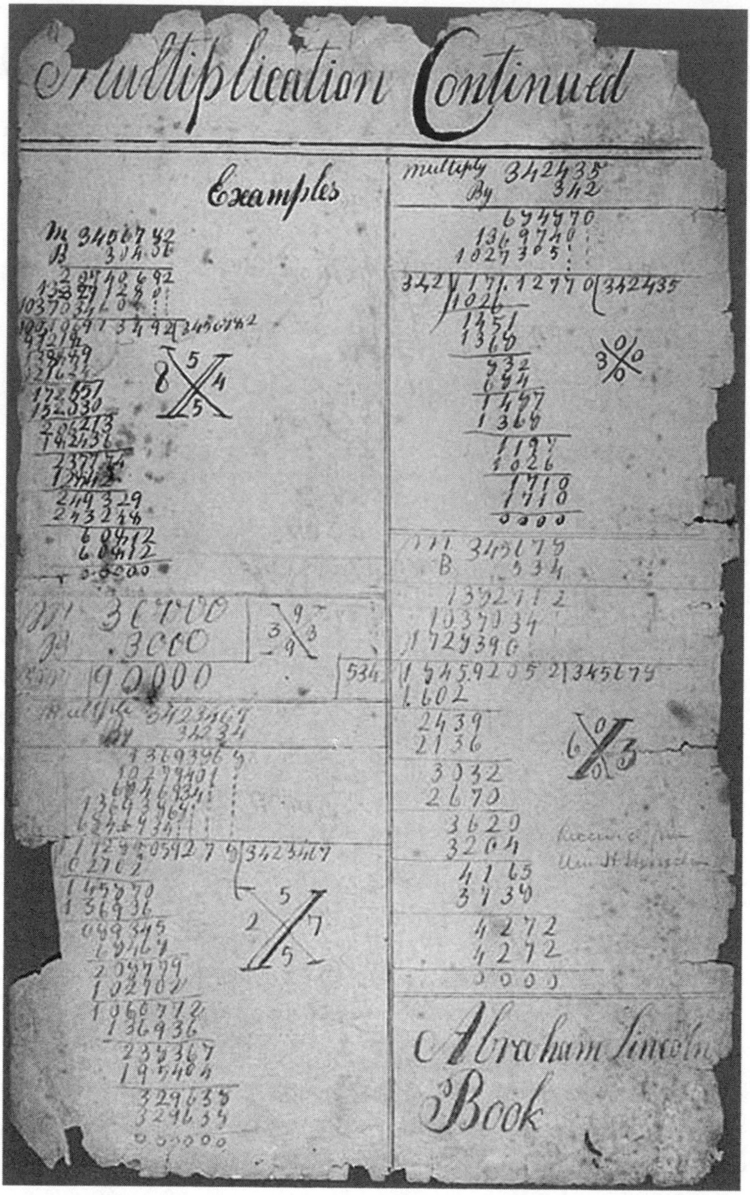

Figure 14.2. Material copied from an arithmetic text while the author was living in Indiana and working as a lawyer. This is the earliest handwriting sample for this individual.

15 Problems from the *Farmer's Almanac*

Cover of the first *Farmer's Almanac* published in the United States. This annual publication—founded in Morristown, New Jersey, by editor David Young and publisher Jacob Mann—has been published continuously since 1818.

The *Farmer's Almanac* was one of the first popularly conceived periodicals to appear in the new country of the United States of America. A yearly publication, it contained scientifically based information on weather, astronomical phenomena, and the tides, and conveyed advice on planting crops, maintaining a garden, preserving foods, and other topics that would be important to agricultural settlers. It was a manual of information, but further, it was also a source of education and entertainment. It is therefore not too surprising to find that it contained mathematical problem sets for the readers to solve. What is surprising is the level of mathematical sophistication one finds within these problems. There were many almanac publishers in the nineteenth century. Here are a few problems from these farmers' almanacs. Some of the phrasing may seem strange, but try to interpret it.

Problems

1. Suppose the area of an equilateral triangle be 600. The sides required.

2. A number is required; that its square shall be equal to twice its cube.

3. There are two numbers whose sum is equal to the difference of their squares, and if the sum of the squares of the two numbers be subtracted from the square of their sums, the remainder will be 60. What are the two numbers?

4. A man filled his store with several sorts of grain, of which $\frac{1}{2}$ was wheat, $\frac{1}{4}$ rye, $\frac{1}{8}$ Indian corn, and $\frac{1}{10}$ oats, and he had 8 barrels of barley. How much of each did he have and what was the total?

5. Four men owned $90 between them, so that if you add X amount, it equals the second man's money diminished by X, the third man's money multiplied by X, and the fourth man's money divided by X. What was each man's part of the $90?

6. A gentleman having a garden 960 feet long and 720 feet wide desired a ditch to be dug about it 5 feet deep; what must be its width, that the earth taken from it may raise the remainder of the garden 1 foot?

7. A man went to market with 100 pounds and bought oxen, sheep, and geese. He gave 5 pounds for each ox, 1 pound for each sheep, and 1 shilling for each goose. The whole number of animals bought was 100; required, the number of each kind.

8. The interest of $800 for one year, and the interest on the interest for another year, is $62.205. What is the rate of interest in percent per annum?

9. What distance does the point which touches the rail on the surface of a railroad wheel 2 feet in diameter pass through while the wheel is going the distance of 62.832 miles?

10. The sum of $500, put out at 5% compound interest, has now amounted to $900; how long has it been at interest?

11. Two men (A and B) bought a pile of pumpkins (1000 in number) for $12; and each paid $6. On dividing them [the pumpkins] it was mutually agreed that A should have his choice of the pumpkins, and allow them to be rated nine mills apiece higher than B's. It is required to divide pumpkins according to the said agreement. How many pumpkins does each receive?

12. There is a cubical block of marble whose superficial contents are equal to 864 times a certain unknown number, and its solid contents equal to 570, 6 times the square of said number. Required, the dimensions.

13. Three men (A, B, and C) where each has certain number of dollars; and A says, "If I had 121 more dollars, I should have as many as both B and C." Says B, "If I had 121 more dollars, I should have twice as many as both A and C." Says C, "If I had 121 more dollars, I should have three times as many as both A and B." How many dollars has each?

14. A man had 10 sheep which he kept until they were 10 years old. They produced a ewe lamb every year, and every one of those lambs and their prosperity when one year old brought forth a ewe lamb. How many were the prosperity of the sheep when 10 years old?

15. A gentleman set out with a certain number of guineas in his pocket but by accident lost 70 of them; but preceding on his journey, he luckily found a purse of dollars, which contained just so many dollars as he had guineas when he set out. At the end of his journey he computed, that, if subsequently he had found half as many guineas as he lost, besides those dollars mentioned before, and next lost one quarter as many guineas as he found, he should then have as many as he had when he set out. Required, his first number. [Assume an exchange rate of $4=1 guinea.]

16.

> There's a plain rectangled Triangular Field
> Which one acre and a half, exactly doth yield.
> Its inscribed Square, has likewise been found
> Just $^{10}/_{16}$ of an Acre of Ground.
> From hence by simple Equation produce
> The base, Cathetus and Hypotenuse.

143

17. Two porters agreed to drink off a quarter of strong beer between them, and two pulls, or a draught, each. Now the first having given it a black eye, as it is called, or drank till the surface of the liquid touched the opposite edge of the bottom, gave the remaining part of it to the other: what was the difference of their shares, supposing the pot was the frustrating of a cone, the depth being 5.7 inches, the diameter at the top 3.7 inches, and that of the bottom 4.23 inches?

18. A's stock was to B's as 3 is to 4. Each man's money was in use as many days as it numbered dollars. A quit with $330 and B with $480. Find each man's stock investment by arithmetical analysis.

19. A little bucket, one-third full, is 8 inches deep, and its upper and lower diameters are 7 inches and 6 inches, respectively. How large is the frog which, jumping into the bucket, causes the water to rise 3 inches?

20. A boy standing on the top of a tower, whose height was 60 feet, observed another boy running towards the foot of the tower, at a rate of 5 mph, on the horizontal plane; at what rate is he approaching the first when he is 80 feet from the foot of the tower?

21. In a mechanic's yard, I observed two wheels 4 feet in diameter so placed that the first was in contact with an upright post, and the second on a level with it, and at a such distance that a line drawn straight from the top of the post to the ground, passing over one and under the other, formed a tangent to both; while a second line from the same point above, passing over the second wheel and touching its periphery, reached the ground 10 feet from the first line. Required, the height of the post.

∾

WHAT ARE THEY DOING?
Popular Problem Situations

Some word problems have become famous in this sense that they are used over and over again at different times and in different societies. Such problems have become a convenient template for conceiving new problems. The social context is changed to conform to the appropriate milieu, but usually the mathematical content considered is the same.

Below is one such series of problems, all of which emanated from the mathematical situation described as the "hundred fowls problem." The problem involves an indeterminant situation that must be solved. This is one of the first such problems in the history of mathematics. Although the problem has existed for over a thousand years, the name "hundred fowls" was attached to it by the mathematical historian and translator of Chinese mathematics books, Louis van Hee (1873–1951) in the early twentieth century.

From *Zhang Qiujian's Mathematical Manual*, China, circa fifth century CE:

A cock is worth five coins; a hen, three coins, and three chicks, one coin. With 100 coins we buy 100 of them. How many cocks, hens, and chicks are there?

From Mahavira, India, circa 850 CE:

Pigeons are sold at the rate of 5 for 3 *panas,* sarasa birds at the rate of 7 for 5 panas, and peacocks at the rate of 3 for 9 panas. A certain man was told to bring at these rates 100 birds for 100 panas for the amusement of the king's son, and was sent to do so. What amount does he give for each of the various kinds of birds that he buys?

From medieval Europe, circa 880:

A merchant wanted to buy 100 pigs for 100 pence. For a boar, he would pay 10 pence, for a sow, 5 pence, while he would pay 1 penny for a couple of piglets. How many boars, sows, and piglets must there have been for him to have paid exactly 100 pence for 100 animals?

From medieval Islam:

A Turkish bath has 30 visitors in a day. The fee for Jews is three *dirhams*, for Christians two dirhams, and for Moslems a half a dirham. Thirty dirhams were earned by the bath. How many Christians, Jews, and Muslims attended?

From sixteenth-century Germany:

Twenty people—men, women, and girls—have drunk 20 pence worth of wine. Each man pays 3 pence, each woman 2 pence, and each girl 1 halfpenny. Required, the number of each.

From Victorian England, 1880:

If 20 men, 40 women, and 50 children receive £350 for seven weeks' work, and two men receive as much as three women or five children, what salary does a woman received for a week's work?

A version that has been repeated in English-language mathematics books since the time of Alcuin, late eighth century:

If 100 bushels of corn be distributed among 100 people in such a manner that each man receives 3 bushels, each woman 2, and each child ½ bushel, how many men, women, and children were there?

As a further exercise, the diligent reader can seek out other variations of this problem in mathematics books.

16 Nineteenth-Century Calculus Problems

Spectators view the future at the 1876 United States Centennial Exhibition held in Philadelphia. This engraving, of the time, shows them marveling at the colossal 700-ton Corliss steam engine, which could produce 1,600 horsepower.

Although by the nineteenth century the United States and Great Britain were well into the Industrial Age, applied calculus problems were rather academic and did not seem to reflect societal needs. However, two of the selected problems below concern the harnessing of steam power, a concession to the new era.

Problems

1. Determine the dimensions of the least isosceles triangle *ACD* that can circumscribe a given circle.

2. Determine the greatest cylinder that can be inscribed in a given cone.

3. Find the greatest value of *y* in the equation $a^4x^2 = (x^2 + y^2)^3$.

4. Determine the different values of *x*, when that of $3x^4 - 28a^3 + 84a^2x^2 - 96a^3x + 48b^4$ becomes a maximum or minimum.

5. Given the fraction $ax/(a-x)$, convert it into an infinite series.

6. Prove geometrically that the hypocycloid is a straight line when the radius of the rolling circle is one-half the radius of the fixed circle.

7. A railway train running at the rate of 30 mph strikes a snowdrift and is brought to a standstill after going 200 yards. Assuming the drift offers a constant resistance to the passage of the train, find how long the train keeps in motion.

8. A copper water tank in the form of a rectangular parallelepiped is to be made. If its length is to be *a* times its breadth, how high should it be that for a given capacity it should cost as little as possible?

9. Show that the curves $x - y = a$ and $2xy = b$ cross at right angles.

10. A man is walking across a bridge at the rate of 4 mph when a boat passes under the bridge immediately below him rowing at 8 mph. The bridge is 20 feet above the boat. How rapidly are the boat and the pedestrian separating 5 minutes after the boat passes under the bridge?

11. A cylindrical tin tomato can is to be made which shall have a given capacity. Find what should be the ratio of the height to the radius of the base that the smallest possible amount of tin shall be required.

12. The cost per hour of running a certain steamboat is proportional to the cube of its velocity in still water. At what speed should it be run to make a trip upstream against a 4 mph current most economically?

13. A rifle ball is fired through a 3-inch plank, the resistance of which causes an unknown constant retardation of its velocity. Its velocity on entering the

plank is 1000 feet/second and on leaving the plank is 500 feet/second. How long does it take the ball to traverse the plank?

14. A woodcutter starts to fell a tree 4 feet in diameter and cuts halfway through. One face of the cut is horizontal, and the other face is inclined to the horizontal at an angle of 45°. Find the volume of the wood cut out.

15. A series of circles have their centers on an equilateral hyperbola and pass through its center. Show that their envelope is a lemniscate.

WHAT ARE THEY DOING?

Graphically Generating a Conic Section, the Parabola

The illustration in figure 16.1 is from Albrecht Durer's *Treatise on Mensuration*, published in 1525. Dürer (1471–1528) was a German artist and

Figure 16.1. Albrecht Dürer's 1525 projection of a parabola as a conic section.

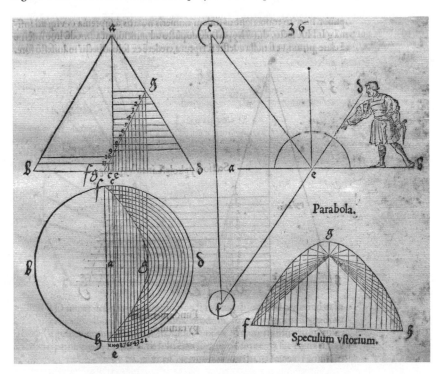

engraver, a mathematician, and an instrument maker. He was particularly interested in the relationships between art and mathematics. He employed projection techniques, foreshadowing the development of descriptive geometry, to explore the intersection of planes and solids. Here on page 33 of his work, he demonstrates that the intersection of a plane with a right cone at an acute angle to the cone's axis produces a parabola. A printing error displaces the resulting parabola. It is askew. The parabola should be rotated 90 degrees clockwise so that points h and f project from the left image to points h and f in the right image.

Such graphical techniques are still used at times today to explore and resolve mathematical issues.

17 Some Sample Problem Solution Methods

Woodblock print from Johann Böschenstein's *Rechenbiechlin* (Augsburg, 1514). The scene shows a person using a slate to work a problem using the "new" Arabic numerals.

In reading the sets of problems in the preceding chapters, you may have wondered, How did they, the original readers, do this? The following case studies illustrate how historical problems would have been solved in their own times by the methods available and how they might be solved by a modern student. Remember that prior to the seventeenth century, mathematics problems were written down in words. The use of algebraic symbolism was a later development. Therefore, the methods used to solve problems in earlier centuries will not necessarily contain algebraic symbolism as we know it.

In the case studies below, where the problem's statement or solution is so complex that it would require an extensive written explanation, I use some convenient symbols or abbreviations. In some cases, I provide a brief explanation concerning the methods used; however, in others I leave the computation for the reader to decipher mathematically. There are also some appropriate questions for the reader to consider.

CASE STUDY 1: ANCIENT BABYLONIA (CA. 2000–1600 BCE)

This problem is found on cuneiform tablet VAT 6598 in the Berlin Museum:

What is the length of the diagonal of a door of height 40 and width 10?

Historical Solution

Babylonian computation was conducted employing a sexagesimal number system (base 60). For ease of understanding, this problem will be solved using the more familiar decimal numeration system.

Recognizing that the width and height form a right triangle, the scribe would apply the reasoning that later came to be known as the "Pythagorean theorem":

$$(40)^2 + (10)^2 = 1700 = (\text{diagonal})^2$$
$$\sqrt{1700} = ?$$

Since the scribe knows that $\sqrt{1700} > 40$, he (all scribes in the ancient world were male) guesses 41, and then computes:

$$1700/41 = 41.463; \quad (41 + 41.463)/2 = 41.23; \quad 41.23 \times 41.23 = 1699.9.$$

And, he concludes, $\sqrt{1700} = 41.23$.

Modern Solution

Nowadays a student (male or female!) would draw a picture of the door, label the given knowns and required unknowns, recognize a Pythagorean situation, and compute $(40)^2 + (10)^2 = 1700$, using a calculator. The diagonal is found to be 41.2311.

CASE STUDY 2: ANCIENT EGYPT

Problems 24 to 27 of the Rhind Mathematical Papyrus (1650 BCE) are concerned with what today would be recognized as obtaining the solutions for simple linear equations. Below is problem 26 of this collection and its historical solution by the method of "false position." Here, the scribe takes a guess at the solution value. The guess is placed into the requirements and an answer obtained. If the answer is not correct, the guess is mathematically manipulated to make it correct.

A quantity and its fourth part becomes 15.

Historical Solution

Given:	Goal:
A quantity and its fourth part	15
Guess 4; then $4 + 4/4 = 5$; $5 \neq 15$.	
5 is not correct, but $5 \times 3 = 15$.	
5 is $1/3$ the correct answer; therefore,	
$(3 \times 4) + ([4 \times 3]/4) = 15$	15
The desired quantity must be 3×4, or 12.	
$12 + 12/4 = 15$	Goal obtained!

Modern Solution

Let x be the unknown quantity; then $x + x/4 = 15$.
Simplifying using algebra, $4x + x = 60 \Rightarrow 5x = 60, x = 12$.

While the "guessing" of a solution to a problem may seem unusual, is it? Can you give examples where the guessing or approximation of an answer would be a part of the solution process for a problem?

CASE STUDY 3: ANCIENT CHINA

The seventh problem of chapter 8 in the *Jiuzhang suanshu* (ca. 100 CE) reads as follows:

Now there are five cattle and two sheep costing 10 *liang* of silver. Two cattle and five sheep cost 8 liang of silver. What is the cost of a cow and a sheep, respectively?

Historical Solution

The Chinese scribe would set up his bamboo computing sticks on the computing board in the following array:

$$
\begin{array}{cc}
2 & 5 \\
5 & 2 \\
8 & 10
\end{array}
$$

He would then multiply the first column by 5 and the second column by 2:

$$
\begin{array}{ccccc}
2\,(5) & 5\,(2) & & 10 & 10 \\
5\,(5) & 2\,(2) & = & 25 & 4 \\
8\,(5) & 10\,(2) & & 40 & 20
\end{array}
$$

Then subtracting the first column's terms from the second, he obtains the configuration

$$
\begin{array}{cc}
0 & 10 \\
21 & 4 \\
20 & 20
\end{array}
$$

From this, he reads that 21 sheep = 20 liang of silver, or that a sheep cost $^{20}\!/_{21}$ liang of silver. Substituting this price into the array in the second column, the scribe finds that

$$10 \text{ cattle} + (^{20}\!/_{21})(4) \text{ liang} = 20 \text{ liang}$$
$$10 \text{ cattle cost } 20 - 4.05 \text{ liang or 1 cow costs 1.52 liang}$$

Can you understand what this scribe did?

Modern Solution

Let cost of a cow be x and the cost of a sheep y:

$$5x + 2y = 10$$
$$2x + 5y = 8$$

Solving this system simultaneously:

$$2(5x + 2y = 10) \Rightarrow 10x + 4y = 20$$
$$5(2x + 5y = 8) \Rightarrow -10x - 25y = -40$$

Adding the two equations together, we obtain $-21y = -20$, or $y = {}^{20}/_{21}$ liang. Substituting this value for y into either equation and solving for x, we find $x = 1.52$ liang. Thus, a sheep costs ${}^{20}/_{21}$ liang and a cow costs 1.52 liang.

CASE STUDY 4: ANCIENT CHINA

Problem 19 in chapter nine of *Jiuzhang suanshu* reads as follows:

A square-walled city of unknown dimensions has gates at the center of each side. It is known that there is a tree 30 *bu* from the north gate, and standing 750 bu from the west gate, one can see the tree. What are the dimensions of the city?

Historical Solution

The Chinese visualized the situation as shown below and obtained a solution using geometric algebra:

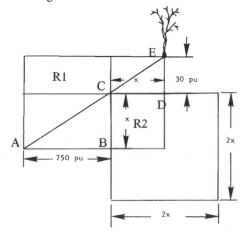

A scribe working on this problem would seek rectangles that contain the knowns and unknown. $R1$ and $R2$ satisfy this condition and are equal in area. Therefore, $(750)(30) = x^2$ and

$$2x = \sqrt{(4)(30)(750)} \text{, or } 2x = 300 \text{ bu, a Chinese mile}$$

Modern Solution

A student would draw the above diagram and note similar triangles containing the given knowns and the required unknown. Triangle $ABC \approx$ triangle CDE. Then the relationship of their sides gives

$$750/x = x/30 \Rightarrow x^2 = (30)(750)$$
$$x = 150, \; 2x = 300 \text{ bu}$$

CASE STUDY 5: ISLAMIC WORLD

This problem is from al-Khwarizmi's *Algebra* (ca. 830):

A square and ten roots are equal to 39 *dirhams*. Find the root.

Historical Solution

Al-Khwarizmi describes the solution technique as shown below. (For the reader's convenience the desired root is represented by x.)

Construct a square representing the square term.
Divide 10 by 2, obtaining 5; adjoin to the square two rectangles $5 \times x$.
Extend their sides to complete the larger resulting square.

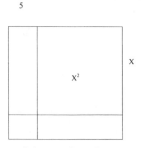

It is known that the areas of the enclosed square and rectangles equal 39 dirhams. Since the remaining corner square has an area of 25, the area of the large square is

$$25 + 30 = 64 \text{ and its side } (5 + x) = 8.$$

Thus, $x = 3$ dirhams.

Modern Solution

The student will recognize that a quadratic equation, $x^2 + 10x - 39 = 0$, is involved here. Compare this with the general formula for a quadratic equation, $ax^2 + bx + c = 0$, and apply the formula for the roots:

$$x = \frac{-b \pm \sqrt{b^2 - 4ac}}{2a}.$$

Thus,

$$x = \frac{-10 \pm \sqrt{100 - (4)(39)}}{2} \to \frac{-10 \pm \sqrt{256}}{2}$$

$$x = (-10 \pm 16)/2 \qquad x = \{3, -13\}$$

This answer obtains both roots. Remember before the seventeenth century, negative roots were not fully understood and were ignored.

CASE STUDY 6: RENAISSANCE ITALY

In the *Treviso Arithmetic* (1478), the first printed arithmetic book in Europe, the author shows his young merchants-in-training how to multiply two multi-digit numbers:

Find the product of 934 and 314.

Historical Solution

At this time, there were seven popular algorithms for doing multiplication. The one shown below, called the *gelosia* method of multiplication, was perhaps the easiest to follow. The student constructed a 3×3 grid of split cells, as shown, and wrote one factor on the top of the grid and the other factor of the multiplication along the right side of the grid.

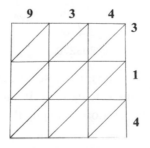

Thus, each column and row corresponded to a single number. The product of the intersecting row and column numbers were then computed and written in the designated cell, with the unit's digits below the bisecting diagonal and the ten's digit above.

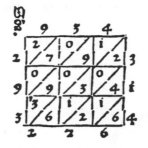

Then the entries along each diagonal beginning at the lower right corner were added. The results were written outside the large square at the lower end of the diagonals.

Reading the product down the left side of the square and along the bottom, we find that

$$934 \times 314 = 293276$$

Can you follow and understand these steps? If so, devise your own multiplication problem and solve it by this method.

Modern Solution

Now, a student would probably use a calculator to obtain the required product, but if not, he or she would most likely use the "downward" method of multiplication.

$$
\begin{array}{r}
934 \\
\times 314 \\
\hline
3736 \\
9340 \\
280200 \\
\hline
293276
\end{array}
$$

This method was also known during the Renaissance but was less popular with the people who did the calculations. What would make one algorithm for a mathematical operation more popular than another?

CASE STUDY 7: RENAISSANCE EUROPE

The following is a fifteenth-century problem requiring use of the "rule of three":

If 100 pounds of sugar are worth 32 *ducats*, what will 9812 pounds be worth?

Historical Solution

Applying the rule of three:

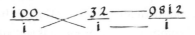

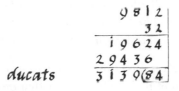

Modern Solution

If 100 pounds are worth 32 ducats, then $^{32}/_{100} = 0.32$ ducats/pound. One then calculates, either through the downward method of multiplication shown in case study 6 or by use of a computer, that $9812 \times 0.32 = 3139.84$ ducats.

Which Is the Better Method?

History testifies that all civilizations were concerned with solving mathematical problems. The problems pondered by our ancestors frequently involved their survival, ensured societal harmony, and prepared the way to the future. Our predecessors determined techniques and various schemes to solve their problems. Whether assisted by an abacus, a knotted string, or a set of counting rods, they obtained their answers, applied them, and moved forward. Mathematical problem solving has a long and fruitful tradition. We are not alone in our efforts.

WHAT ARE THEY DOING?

Problems of Partnership from the Renaissance

The two-page problem shown in figure 17.1 is from the *Treviso Arithmetic* (1478), the first printed arithmetic book in Europe. It is written in Italian. The problem involves partnership: the distribution of costs and profits. This was an important topic of consideration in early arithmetic books that focused on the needs of trade and commerce. The problem is stated as follows:

Three merchants, viz., Piero, Polo, and Zuanne, went into partnership. Piero put in 112 ducats, Polo 200 ducats, and Zuanne 142 ducats. They found that they had gained 563 ducats. Required, the share of each.

Then the writer gives directions on how to solve the problem. First, the contribution of each of the members is written down and added to arrive at the sum of 454 ducats. He calls this number the divisor. Then he considers the case of Piero, saying "If 454 ducats gained me 563 ducats, how much should 112 ducats gain me?" He then solves this proportion by the rule of three, using the first diagram. The product of 563 and 112 is found by the "chessboard method," an algorithm for multiplication that we recognize as the one used today. The product 63,056 is obtained.

Next, this number is divided by 454 using the "galley method" of division. The quotient is found to be 138 ducats with a remainder of 404. At this period of history 1 *ducat* = 24 *grossi* and 1 *grossi* = 32 *picoli*. The remainder, 404 is changed into 9696 grossi and then divided by 454. The result of the second division gives us 21 grossi and a remainder of 162. The 162 grossi are changed into 5184 picoli, and this number is divided by 454. The result of the computation is that Piero's profit is 138 ducats, 21 grossi, and 11 $^{190}/_{454}$ picoli.

A similar series of computations is demonstrated to obtain the shares for the remaining partners.

La pꝛima raꞇone ſe forma coſi.

Ꞇre merchadāti ꞇoe. Piero Polo e çuanne hāno
fatto compagnia. Piero miſſe ꝺucati.i i 2. Polo
miſſe ꝺucati.2 o o. çuanne nuſſe ꝺucati.i 4 2. Et
hāno trouato ꝺe guadagno ꝺucati.5 6 3. domando
che tocha per homo.
Jn queſta e ciaſchaduna raꞇone ꝺi compagnia tu
metterai tuti li cauedali vno apꝛeſſo laltro.e farai
vna ſomma ꝺe quelli iongēdoli inſieme : la quale
ſara lo tuo partitoꝛe,a queſto modo.

Piero miſſe ꝺucati .i i 2.
Polo miſſe ꝺucati .2 o o.
çuāne miſſe ꝺucati .i 4 2.
   LaꝊomma. 4 5 4. partitoꝛe

Poi metterai la raꞇone ꝺe Piero in forma ꝺicēd●
Ꝩe ꝺucati .454. me guaꝺagniuo ꝺuccati.5 6 3.
che me guadagueruo ꝺucati.i i 2. Oꝛ tu cogno
ſci il to partitoꝛe.mettuda adoncha la regula i foꝛ
ma:tu ſai che ꝺei fare:ſe gōdo li comādamēti ſoi.
Unde metti la regula in foꝛma coſi.

Et e fatta. Unde reſpoudi che tochara a Piero ꝺe
guadaguo ꝺucati 138 g̅ 21 p̅ i i.e   1 9 0
             4 5 4

Figure 17.1. Dividing up profits in the fifteenth century, an example of partnership
problems.

18 Where to from Here? Where Do You Want to Go?

Woodblock print from Petrus Apianus's *Introductio Geographia* (1533/1534), the first book to consider the use of applications of trigonometry in the study of geography.

The intent of this book is very specific and limited: to encourage the use of historical mathematics problems in classroom instruction. Nevertheless, word problems themselves—in their appeal, interpretation, understanding, and solution processes—constitute a broad and complex subject that affects mathematics teaching and learning. There are special difficulties involved in teaching problem solving via word problems, from the reading abilities of the students in a class, to their organizational skills, to the cultural and economic relevance of the situations discussed. Each of these topics deserves attention. I once gave a series of consumer-oriented problems to students involving the purchase of fuel oil only to realize later, to my chagrin, that these students were from poor families that could not afford oil. They heated their houses with coal.

Fortunately, there is a rich and varied literature that addresses these concerns extending from such basic works on problem solving as George Polya's classic, *How to Solve It* (1971), or the more teacher-friendly *Creative Problem Solving in School Mathematics*, by George Lenchner (2005), to more recent, comprehensive compilations of opinion and research such as *Words and Worlds: Modeling Verbal Descriptions of Situations* by Lieven Verschaffel and colleagues (2009). Articles in teachers' magazines and journals frequently discuss issues of problem solving.

The illustrative problems assembled in this book are intended for general classroom instruction; therefore, the rigor of the problem-solving situations was constrained. Collections of more demanding "challenge problems"— particularly those employed in various national and international mathematics contests and competitions—are readily available. Students of high ability should have a chance to try such problems. Readers are encouraged to familiarize themselves with this literature to improve their understanding of problem solving in general. Now let us return to the topic of employing historical problems in classroom teaching.

The problems selected and offered in this book are but a mere sample of the thousands of student exercises that have been posed since the beginning of formalized mathematics instruction. Readers seeking fuller collections of problems are urged to explore the contents of the references given in the bibliography. Large libraries such as those associated with a university may contain old mathematics texts from which problems can be mined. Occasionally, bookstores specializing in old books will have mathematics texts for sale. I have purchased several early American mathematics texts in garage sales and flea markets. The Internet is also a rich and easy means for obtaining old mathematics problems; in particular, websites such as MathDL (the MMA digital library; mathdl.mma.org) and The Math Forum (mathforum.org) contain collections of problems suitable for use in middle and secondary school and university mathematics teaching.

The learning and versatility of historical mathematics problems should be further explored through discussions and assignments, including research projects. For example, after students have considered and solved a series of problems from a specific historical period or geographical region, their attention could be directed to discussion questions such as, "What could you tell about the life of these people?" "What were their occupations?" "What was their monetary system?" From early problems, we can learn about the human dependence on agriculture: the growing, storing, transportation, and distribution of grain. Later historical periods and their problems testify to the rise of trade and the development of a sophisticated mercantile mathematics that introduced such concepts as percentage, profit, loss, commissions, tolls, and taxes. The rise of European trade from the Middle Ages onward spawned voyages of exploration; how do the content of problem situations testify to these movements? "What products were imported?" "Exported?" "How did the commodity of pepper influenced European trade?" "Why was such a spice important at the time?" "Where did these spices come from?" "Who controlled the spice trade?"—and the expedition of inquiry can go on. Probing another historical development, "How is warfare represented in problems?" "Did warfare effect mathematics?" "Does it today?" Much of our understanding of the properties of a parabola is due to Renaissance research of artillery trajectories (Swetz 1989).

Through a consideration of the contents of historical mathematics problems, windows on human history are opened, allowing a broader, clearer, more perceptive view of the landscape. The evolution of measurement standards and the development of monetary systems are revealed. Historical mathematics problems provide a strong link to interdisciplinary teaching. Topics considered in world studies classes can be coordinated with the appropriate historical problems. A student project I have found successful on several occasions is, after encountering Egyptian problems that reflect on the construction of pyramids, to have students investigate pyramids in other societies. "Did you realize the ancient Babylonians had pyramid-like structures, *ziggurats,* as did the Chinese, the Inca and the Aztecs?" "What mathematical similarities or differences did their pyramids have?" "Were the great mounds built by the Mississippian culture really earthen pyramids?" And the exploration can go on. I sincerely hope that in the classrooms of the teachers who have read this book, such adventures take place!

Acknowledgments

Any written work uses contributions and influences from many sources. I would like to recognize persons and institutions that were especially helpful in compiling this book. Historical materials such as engravings and woodblock prints that are in excess of 100 years old are in the public domain unless there are unique circumstances relating to their existence and use. If such circumstances exist, I have attempted to respect them. Where possible, I have identified the historical origins of the items used. Special courtesies were extended in the securing of permissions for use of the following:

Frontispiece, Shipwreck scene: John Heilbron, *Geometry Civilized*, frontispiece.
Page x, Euler's frontispiece: "Mathematical Treasures" from the Convergence section of *Loci,* Mathematical Association of America (MAA) Mathematical Sciences Digital Library website, http://mathdl.maa.org/mathDL, and the Plimpton-Smith Collection, Columbia University Libraries.
Chapter 1. I would like to thank Michel Lokhorst and Sense Publishers, Rotterdam, for allowing me to republish my contribution to the book, "Word Problems: Footprints from the History of Mathematics," pp. 73–93 in *Words and Worlds: Modeling Verbal Descriptions of Situations*, ed. L. Verschaffel et al. (2009).
Chapter 1, Kepler illustration: "Johann Kepler's *Astronomia Nova*," from the Convergence section of *Loci,* March 2010, MAA Mathematical Sciences Digital Library website, http://mathdl.maa.org/mathDL.
Chapter 2, figure 2.4: Plimpton-Smith Collection, Columbia University Libraries.
Chapter 3, p. 38, YBC 7289: Photo courtesy of Ulla Kasten, Yale Babylonian Collection, Yale University.
Chapter 3, figure 3.1a, sketch of YBC 7289: ©Mathematical Association of America 2011. All rights reserved.

Chapter 4, p. 46, Egyptian scene: Photo from archive of Jon Bodsworth.

Chapter 7, p. 77, Lilavata illustrations in problems 19 and 20: Plimpton-Smith Collection, Columbia University Libraries.

Chapter 8, p. 80, Arab scholars: *Aramco World*.

Chapter 9, problems 30–38: translation from Levi ben Gershon, *Maaseh hoshev* (The art of calculation) by Shai Simonson, Stonehill College, MA.

Chapter 11, problems 18, 19, 22, 30: Taken from *Sacred Mathematics: Japanese Temple Geometry,* by Fukagawa Hidetoshi and Tony Rothman (Princeton University Press, 2008), with permission of the authors and Princeton University Press.

Chapter 11, pp. 105, 107, 108, 110, and 113, illustrations of pouring of oil, finding the missing numbers, multiplying mice, dividing cloth under a bridge, and the stack of barrels: Personal collection of Fukagawa Hidetoshi.

Chapter 13, p. 122, Lewis Carroll portrait: Plimpton-Smith Collection, Columbia University Libraries.

Chapter 14, figure 14.1, page from the Danner cipher book: Image courtesy of The Hershey Story, The Museum on Chocolate Avenue.

Chapter 14, figure 14.2, Lincoln page: Plimpton Collection, Rare Book and Manuscript Library, Columbia University Libraries.

Chapter 16, figure 16.1, Dürer's illustration: Plimpton-Smith Collection, Columbia University Libraries.

I would also like to thank Sara Mrljak for her assistance in the typing of the final manuscript and Greg Crawford and the staff of the Heindel Library, Penn State Harrisburg, for their diligence in securing often obscure research materials. Greg Nicholl of the Johns Hopkins University Press supplied guidance on manuscript preparation, and Steve Swetz assisted in the processing of images. Finally, a special thanks is due to my wife, Joan, who proofread the final draft, and to Carolyn Moser, who supplied the critical final reading and made many useful suggestions for the improvement of the final manuscript.

Answers to Numbered Problems

Chapter 3. Ancient Babylonia

1: 18 units. **2:** 11.7 units. **3:** 3.05 units. **4:** 30 and 25 units. **5:** 31.25 units. **6:** 32 and 24 units. **7:** 13 cubits. **8:** 4.375 shekels and 4.125 shekels. **9:** 171.428 square units and 128.57 square units. **10:** 2500 laborers; 833$\frac{1}{3}$ baskets of barley. **11:** top level, $\frac{1}{6}$ day; middle level, $\frac{1}{3}$ day; bottom level, $\frac{1}{2}$ day. **12:** The granary will pay for 16,4571 man-days of work. **13:** $\frac{1}{2}$ unit. **14:** 180 units. **15:** 1.6 shekels. **16:** 4.5 cubic units. **17:** 4 months, 24 days. **18:** 12 rods. **19:** $\frac{1}{2}$ unit. **20:** 30 units. **21:** 30 units and 25 units. **22:** $\frac{1}{6}$ unit. **23:** 30 units and 20 units. **24:** 30 units and 20 units. **25:** length, 5 rods; width, 1.5 rods; depth, 6 rods. **26:** area, 7.5 square rods; volume, 3.7 cubic rods; workers, 22.5 or 23; wages, 138 sila. **27:** 1944 sar. **28:** A trapezoid is used as an approximation. **29:** 900 square cubits. **30:** 773 square cubits.

Chapter 4. Ancient Egypt

1: 14.28 units. **3:** normal share, 7$\frac{9}{13}$ loaves; special share, 15$\frac{5}{13}$ loaves. **4:** 315. **5:** 40 hekats, 26$\frac{2}{3}$ hekats, 20 hekats, and 13$\frac{1}{3}$ hekats. **6:** 29.02 square units. **7:** 340.34 cubic cubits. **8:** 12 cubits and 5 cubits. **9:** 8 cubits and 2 cubits. **10:** 25.82 cubits. **11:** 16$\frac{5}{8}$. **12:** 9. **13:** 16.0206. **14:** 6 cubits and 8 cubits. **15:** The shares are as follows: $\frac{7}{16}$, $\frac{9}{16}$, $\frac{11}{16}$, $\frac{13}{16}$, $\frac{15}{16}$, 1$\frac{1}{16}$, 1$\frac{3}{16}$, 1$\frac{5}{16}$, 1$\frac{7}{16}$, and 1$\frac{9}{16}$. **16:** $1 + \frac{1}{2} + 16$, $10 + \frac{1}{2} + \frac{1}{3}$, 20, $29 + 16$, and $38a + \frac{1}{3}$. **17:** 785.4 cubic cubits. **18:** The area of a circle is approximated by squaring $\frac{8}{9}$ its diameter; $\pi = 3.24$. **19:** 1050, 1200, 2550, 5100. **20:** 390.5 cubits. **21:** 94.23 cubits. **22:** $\frac{5}{6}$ cubits. **23:** 5$\frac{1}{25}$ hands. **24:** 3$\frac{1}{3}$ pairs per day. **25:** 20 jars.

Chapter 5. Ancient Greece

1: 8 and 12. **2:** 4, 7, 9, 11. **5:** 13 and 3. **6:** 0.48 days. **7:** 2$\frac{6}{7}$ and 4$\frac{2}{7}$. **8:** 8 and 12. **9:** 4, 7, 9, 11. **10:** 60 years old. **11:** $\frac{2}{5}$ of a day. **12:** 3$\frac{5}{7}$ and 4$\frac{6}{7}$. **13:** 45, 37$\frac{1}{2}$, 22$\frac{1}{2}$.

14: 120 apples. **15:** his childhood, 14 years; his youth, 7 years; his marriage, at age 33; birth of his son, at age 38; his son dies, at age 80; his period of grief, 4 years; Diophantus's age at his death, 84 years. **16:** Each Grace had $4k$ apples, gave away $3k$ apples, and kept k. If we let $k = 1$, then each Grace and each Muse had one apple. **17:** $144/37$ hours. **18:** 960. **19:** $289/4$ and $49/4$ when $k = 5$. **23:** They are in the ratio of 1:2:3. **26:** 8.944 square units. **31:** One possibility: 9, 328, and 73.

Chapter 6. Ancient China

1: 26 inches. **2:** The one who paid 5 coins receives 2 coins back; the 3-coin payer receives 1.2 coins, and the last payer, 0.8 coins. **3:** hound, 15.857 coins; raccoon, 31.714 coins; fox, 63.429 coins. **4:** 12. **5:** silver, 1.828 units; gold, 2.234 units. **6:** $2^2/17$ days. **7:** 6 feet × 8 feet. **8:** 12 feet 1inch. **9:** military horse, 22.857 dan; ordinary horse, 17.143 dan; inferior horse, 5.714 dan. **10:** 15.7068 days; good horse, 4534.2408 li; inferior horse; 1465.7592 li. **11:** square, 239.682 bu; pond, 60.318 bu. **12:** no; area = 759.57 square paces. **13:** one possible value: 233. **14:** 12 rabbits and 23 pheasants. **15:** 8 animals and 7 birds. **16:** 60 days. **17:** first day, 352.755 miles; second day, 176.777; third, 88.188; fourth, 44.094; fifth, 22.047; sixth, 11.023; seventh, 5.512. **18:** 780 li. **19:** 60 guests. **20:** Each side is 252 bu long. **21:** 332, 253, 253, and 162, respectively. **22:** 1.78125 dou. **23:** The pond is 5.6 feet deep; the reeds are 8.6 feet and 6.6 feet long, respectively. **24:** 2.5 days. **25:** the slow walker, 10.5 bu; the fast walker, 24.5 bu. **26:** 28.142 inches. **27:** 29 feet. **28:** cow, $2^6/7$ baskets; horse, $1^3/7$ baskets; goat, $5/7$ baskets. **29:** The height of the island is 4 li 55 pu. It is 102 li 150 pu from the front pole. **30:** 238 feet. **31:** $4^{11}/20$ chi. **32:** depth of the water, 12 chi; length of the plant, 13 chi.

Chapter 7. India

1: 59. **2:** one answer: asavas horses, 42; hayas horses, 28; camels, 24; total value, 262. **3:** $3^1/7$ yojanas. **4:** 5. **5:** 45 pearls. **6:** 10 units. **7:** 89.94 inches or 7.5 feet. **8:** $7/2$. **9:** 540 coins. **10:** altitude, 8; area, 36. **11:** 18 mangoes. **12:** 16 or 48. **13:** $3^1/2$ feet and $22^1/2$ feet. **14:** 4.38 days. **15:** citron, 8; wood apple, 5. **16:** 72 bees. **17:** 317. **18:** 100 panas. **19:** They meet at 12 hastas from the pole. **20:** 50. **21:** altitude, 12; base segments, 5 and 9; area, 84 square units. **22:** 50 coins. **23:** ruby, $47/24$; emerald, $47/30$; pearl, $47/300$. **24:** upright side, 80; base, 60. **25:** area of the field, 108 square units; altitude, 12; diagonal, 15.24. **26:** 50 or 5. **27:** 100 arrows.

Chapter 8. Islam

1: 2.57066, 3.26993, 4.15941. **2:** 5. **3:** One division: 6.345 and 3.655. **4:** 4.26 and 5.74. **5:** 6.0842 and 3.91577. **6:** 6 and 4. **7:** 5.0903. **8:** 4 .8 yards. **9:** 4 dirhams and 2 dirhams. **10:** 5. **11:** husband, $1/4$; son, $3/10$; daughters, $3/20$ each. **12:** husband,

0.183036; son, 0.219644; daughters, 0.10982 each. **13:** 12. **14:** 2 and 8. **15:** 3. **16:** 3.464. **17:** 7 and 3. **18:** 24. **19:** $1\frac{1}{14}$. **20:** 2 dirhams. **21:** 3.0306 and 6.9694. **22:** 6.0842 and 3.9158.

Chapter 9. Medieval Europe

1: It can't! **2:** 1,572 days. **3:** 100^2 nests, 100^3 eggs, 100^4 birds. **4:** 3 feet will extend beyond the top of the smaller pole. **5:** 1, 3, 9, and 27. **6:** 246 years 210 days. **7:** The answer is incorrect. The integral value of the area enclosed is 523 of the house areas, but this is a "packing problem." The houses must be fitted into the available space. The best geometric arrangement allows for 519 houses. **8:** 28. **9:** 1 boar, 9 sows, and 90 piglets. **10:** 9 ounces of gold; 2 pounds 3 ounces of silver; 6 pounds 9 ounces of brass; and 20 pounds 3 ounces of lead. **11:** 3600 sextarii flow in by the first pipe, 2400 through the second pipe, and 1200 through the third. **12:** 400. **13:** Give the first son the 10 half-full flasks, then to the second son 5 full and 5 empty flasks, and similarly to the third son. **14:** 1,073,741,824 men. **15:** None: since the ox pulls a plow, he covers up his footprints. **16:** The men who first spoke had four oxen, and the one who was asked had eight. **17:** Some assumptions must be made:

1. Each brother is attracted to only one other woman in the group.

2. For a transgression to occur, each endangered couple must be alone.

Let the brother-sister couples be represented by B_1,S_1; B_2,S_2; and B_3,S_3. Let the attractions be $B_1{\rightarrow}S_2$; $B_2{\rightarrow}S_3$; and $B_3{\rightarrow}S_1$. Then B_1B_2 cross over. B_2 returns and fetches S_1, crosses back with her, and leaves S_1 with B_1. B_2 now returns and fetches his sister S_2. B_2 remains with S_2 and S_1, and B_1 fetches S_3. Now, B_1,S_1 and B_2,S_2 remain together; and S_3 goes back and retrieves her brother. There are other ways to do this crossing problem. **18:** The two children cross the river, and one brings the boat back. The father crosses in the boat, and the other child brings the boat back. The child fetches his sibling and brings him or her across. Then one of them returns to fetch the mother and crosses again to bring his sibling back to the other shore. **19:** 1000. **20:** 7 acres. **21:** 210 casks. **22:** 3 men, 5 women, and 22 children. **23:** each master builder, 4.54 pence; the apprentice, 2.27 pence. **24:** 19 camels, 1 ass, and 80 sheep. **25:** $1\frac{32}{100}$ bezants. **26:** 6.78 and 8.37 pounds, respectively. **27:** worked, $13\frac{7}{11}$ days; not worked, $16\frac{4}{11}$ days. **28:** corn, $26\frac{7}{10}$; barley, $26\frac{7}{10}$; millet, $12\frac{2}{10}$; beans, $12\frac{2}{10}$; lentils, $12\frac{2}{10}$. **29:** first horse, 12 bezants; second horse, 14 bezants; first man, 8 bezants; second man, 12 bezants. **30:** 13.483. **31:** $64\frac{17}{131}$. **32:** $9\frac{6}{11}$ dinars. **33:** Approximately 6 hours 25 minutes 43 seconds. **34:** $\frac{2}{7}$ of the expensive drug, $\frac{5}{7}$ of the cheap drug. **35:** He gained, with a profit of 103.366 dinars. **36:** 11.5 and 1.5. **37:** first number, 14.34; second number, 14.26. **38:** 178.3166 and 58.8.

Chapter 10. Renaissance Europe

1: 18 units. **2:** 9.615 and 0.385. **3:** 35.09 days. **4:** 47 pieces; shares are 33, 13, and 7. **5:** price, 120 francs; duty,10 francs. **7:** 0.344 scudi/pound. **8:** 2 square units. **9:** R^2 $(\pi - 2)$. **10:** 28 beggars; 220 pennies. **11:** 10. **12:** 301. **13:** fee, 7.74 denarii; price per fish. 21.82 denarii. **14:** base, 12; hypotenuse, 13. **15:** 5 and 8. **16:** 6.8756. **17:** 8 allotments per day. **18:** 7 days. **19:** first man, 33 denarii; second man, 76 denarii; third man, 65 denarii; fourth man, 46 denarii; purse 119 denarii. **20:** 3 feet will extend beyond the top of the smaller pole. **21:** Four journeys will be required, and 20 measures of grain will get through. **22:** 9.8235 and 7.1176 denari. **23:** 0.833 braccia. **24:** ratio/proportions, 0.307, 0.269, 0.242, 0.181; shares, 254.81, 223.27, 200.86, 150.23. **25:** 62.6 days,7.8 braccia, and 46.956 braccia. **26:** 4.615, 12.308, 6.154. **27:** No; the man who gave three loaves should have received 4 bezants, and the other man, 1 bezant. **28:** 1.62 hours. **29:** 12% profit. **30:** five measures of cheap wine with two measures of expensive wine. **31:** 10½ denarii. **32:** 12 denarii at each fair. **33:** 1,048,576 denarii. **34:** 800 ducats, 2.76 grossi. **35:** 9, 16, and 13 denarii, respectively. **36:** Tommaso, 1052 ducats, 11.273 grossi; Domenego, 942 ducats, 3.674 grossi; Nicolo, 1173 ducats, 22.55 grossi. **37:** 16% plus. **38:** 15.34 fl/hundredweight. **39:** 44.3 lires and 55.7 lires, respectively. **40:** 6 months. **41:** 7.966 and 15.93 units. **42:** 20 units.

Chapter 11. Japanese Temple Problems

1: diameter, 9 units, and side, 4 units; or diameter, 7.8242, and side, 0.6793. **2:** $n = 537$. **3:** 7789. **4:** $n = 4$; price $= 60$ yen. **5:** 59 years old. **6:** $r = \dfrac{ab}{2c + 3(a+b)}$. **7:** Using Heron's formula, $r = \sqrt{\dfrac{(AD)(BE)(CF)}{AD + BE + CF}}$. **9:** Let R equal the radii of the sphere and r, the radii of the covering circles. Consider the regular polygons and polyhedrons P_n that can be inscribed in a sphere and meet the following conditions:

1. The vertices of P_n will determine the centers of the covering circles $O_i(r)$ on the sphere's surface.
2. Each face possessed of P_n is an equilateral triangle.

The following values of n will then prove to be the desired solutions:

$$n = 3: P_n \text{ is an equilateral triangle and } r = \frac{\sqrt{3}}{2} R$$

$$n = 4: P_n \text{ is a tetrahedron and } r = \frac{2}{\sqrt{6}} R$$

$n = 6$: P_n is a octahedron and $r = \dfrac{\sqrt{2}}{2} R$

$n = 12$: P_n is a isocahedron and $r = \sqrt{\dfrac{5 - \sqrt{5}}{10}} R$

10: area of pentagon, $3.847a^2$; area of n-gon: $\dfrac{na^2}{4} \cot \dfrac{180°}{n}$. **12:** 3.932 and 5.796.

13: $a = \dfrac{BC(AC - 2b)}{2(AC - b)}$. **14:** 10.1 times. **15:** 2.43 measures. **16:** Let the tub be T, the small ladle l_1, and the large ladle l_2. Then,

1. With l_1, take 3 scoops out of T and fill l_2. Then 1 sho of oil is left in T, with 7 sho in l_2, and 2 sho in l_1.
2. Now pour the contents of l_2 into T. Then we have 8 sho of oil in T, 2 sho in l_1, and nothing in l_2.
3. Pour the contents of l_1 into l_2. There are now 8 sho in T, no oil in l_1. and 2 sho in l_2.
4. Use l_1 to scoop 3 sho of oil from T into l_2. Hence, there are 5 sho of oil in T and 5 sho in l_2, and the customer can be accommodated.

17: (a) 27,682,574,402; (b) $2 \times 7^{12} \times 12$ cm (seven times the distance from the earth to the moon). **21:** 15 thieves, 113 tan. **22:** area, 84.87 square units; side, approx. 4. **23:** 60 days. **24:** $R = 3$. **25:** $5\frac{5}{17}$ days. **26:** 24 days. **27:** $R = (1 + \sqrt{8}) r$.
28: $[(120 + 84)/2] \times 6$. **29:** $t = \dfrac{\sqrt{2r}}{\sqrt{2} + 1} = 0.585785r$. **30:** First-class home, 2.5 koku of rice, and all first-class homes, 10; second-class home, 2 koku, and all second-class homes, 16; third-class home, 1.6 koku, and all third-class homes, 24; fourth-class home, 1.28 koku, and all fourth-class homes, 52.48; fifth-class home, 1.024 koku , and all fifth-class homes 122.88. **31:** $N = 2060$. **32:** The depth of the pond is 12 syaku. **33:** One possible solution is the following:

2	1	3
3	2	1
1	3	2

34: $t = 0.464r$. **35:** $r = 3$. **36:** 1,073,741,823 mon. **37:** 238.24 ri. **38:** 25. **39:** Use summation formula for arithmetic progression: there are 468 barrels.

Chapter 12. The Ladies Diary

1: approx. 31 acres. 2: £5 13s 2¼ pence (where 12 pence = 1 shilling; 20 shillings = £1). 3: Determining how much was ground by each man off the original radius of the wheel, we find: first man, 2.225 inches; second, 2.439 inches; third, 2.658 inches; fourth, 3.038 inches; fifth, 3.604 inches; and the sixth man is left with a grinding stone approximately 4 feet 3 inches in diameter. 5: $2\pi h$. 6: 1.5 yards. 7: 17.02 feet. 8: 163,350 square feet. 9: 17,310,309,456,440 farthings. 10: 1743 yards, or almost a mile. 12: 10 yards. 13: width = $[(a+b) \pm \sqrt{(a^2+b^2)}\,]/2$. 15: £3 1 shilling ½ pence. 16: Approximately 189 gal. 17: Water. 20: One answer: 13 geese at 1 shilling 8 pence each, 11 ducks at 10 pence each, and 6 chickens at 5 pence each. 22: A horseshoe.

Chapter 13. Nineteenth-Century Victorian Problems

1: 529.48 yards. 2: 36. 3: vertical height, 850 yards; distance from first observer, 1482 yards; distance from second observer, 946.5 yards. 4: To the height of its diameter. 5: 162.03 square units. 6: 10%. 7: 19. 8: 60 yards × 60.5 yards. 9: 17.02 feet. 10: 10 yards. 12: cliff, 250 feet; tower, 183 feet. 14: 7381 miles—but there is a mistake! To give the correct answer, the measurement on the middle pole should be 15.987 inches, for a radius of 3963 miles. 15: 20 days. 16: 3 coins. 17: 4 and 6 mph. 18: 60 years and 40 years. 19: 503 in base 7. 20: 35 miles. 21: Let the sides of the right triangle be a, b, and c, where a is the perpendicular side, b is the base, and c is the hypotenuse. Let p be the perimeter. Then it can be found that

$$c = [(p-a)^2 + a^2]/2\,(p-a) \text{ and}$$
$$b = [(p-a)^2 - a^2]/2\,(p-a).$$

22: 1480 pounds. 23: £0.5 per week. 24: Portsmouth, 200; Plymouth, 189; Sheerness, 101. 25: 65 pumps. 26: six horses. 27: two days. 28: £659.29 and £283.71. 29: 7.966 and 15.93. 30: 20 units.

Chapter 14. Eighteenth- and Nineteenth-Century American Problems

1: A, 70 hhds; B, 30 hhds. 2: 3.999 pounds. 3: 76.78 yards or ~230 feet. 4: 135. 5: 8.2 pounds. 6: $S = ab/(a+b)$. 7: $S_1 = 2(4b^2 - a^2/15)^{1/2}$; $S_2 = 2(b^2 - 4a^2/15)^{1/2}$. 8: He lost by $14. 9: cup = 16 oz, cover = 32 oz. 10: $2600. 11: $14. 12: 3888. 13: 12⅚. 14: 11.701 feet. 15: True. 16: 2 pounds of the 8-cent sugar, 2 pounds of the 10-cent sugar, and 6 pounds of the 14-cent sugar. 17: 301. 18: Let the base of the triangle be represented by b and the sum of the perpendicular side, a, and the hypotenuse, c, by k. Then,

$$c = (b^2 + k^2)/2k \text{ and}$$
$$a = (k^2 - b^2)/2k.$$

19: 50. **20:** An ellipse. **21:** 7939.265 miles. **22:** 102.64 feet. **23:** 0 or ½. **24:** 160 barrels of wheat, 80 barrels of rye, 40 barrels of corn, 32 barrels of oats, plus the 8 barrels of barley, to total 320 barrels of grain. **25:** 37.22. **26:** 47.79 inches. **27:** £18, £22, £10, and £40. **28:** 30 and 36. **29:** approx. 33 minutes. **30:** 5.805 square inches. **31:** circle, 0.07958 square meters; square, 0.0626 square meters; triangle, 0.04810 square meters. **32:** 9:4. **33:** 150 miles. **34:** 22.4 pounds. **35:** ¾ and ¼. **36:** 314,146,179 bushels. **37:** He gained $1,500,873. **38:** length, 49 rods; breadth, 1 rod. **39:** 80,120. **40:** 38.1727°. **41:** 3.5972845. **42:** 572. **43:** line, 7.07 chains; distance on first side, 10.61 chains. **44:** See answer for problem 18 in chapter 9. **45:** 15 and 45 years. **46:** 33, 57, and 45 inches. **47:** 4:21.49. **48:** 72. **49:** 22.5 mph and 18.409 mph. **50:** 40. **51:** 60. **52:** September 9. **53:** John, $25,000; Katie, $15,000. **54:** X, $18.98; Y, $17.02; Z, $54.00. **55:** 7 calves, 21 goats, and 72 pigs. **56:** upper section, 17 826 inches; middle section, 4.633 inches; lower section, 3.250 inches. **57:** 8. **58:** 5100. **59:** 6 units. **60:** 15 × 15 × 15 inches. **61:** side = 31.1769 inches. **62:** ¹⁄₆₀ day. **63:** 9 weeks. **64:** $3700. **65:** C paid $360, A received $1800, and B received $360. **66:** A paid $31.03; O, $48.28; and P, $20.69. **67:** James, $0.35 and Henry, $0.05. **68:** Seven peaches were stolen. The first hoodlum received two; the second hoodlum, two; the third hoodlum, two; and the original thief, one. **69:** 3 gallons of $1.10 wine, 4 gallons of $1.80 wine, 15 gallons of $2.50 wine, and 10 gallons of water. **70:** 58 feet. **71:** 696⅔ bushels. **72:** 37. **73:** 36 days. **74:** first horse, $72; second horse, $120.

Chapter 15. Problems from the Farmer's Almanac

1: 37.22. **2:** 0 or ½. **3:** 30.5 and 29.5. **4:** 160 barrels of wheat, 80 barrels of rye, 40 barrels of corn, 32 barrels of oats, plus the 8 barrels of barley, to total 320 barrels of grain. **5:** $18, $22, $10, and $40. **6:** 1.95 feet. **7:** 19 oxen, 1 sheep, and 80 geese. **8:** 7.24%. **9:** 193.39 miles. **10:** 12.05 years. **11:** A, 345 pumpkins; B, 655 pumpkins. **12:** a cube with a side of 4 units. **13:** A, $11; B, $55; C, $77. **14:** 10 $(2^{10} - 1)$ lambs. **15:** 175 guineas. **16:** 1115.3 feet, 117.2 feet, and 1121.44 feet, respectively. **17:** 7.05 cubic inches. **18:** A, $202.50; B, $270. **19:** approx. 4.05 pounds. **20:** 6.25 mph. **21:** 23.43 feet.

Chapter 16. Nineteenth-Century Calculus Problems

1: If the radius of the circle is r, then it is found that the sides of the triangle are $r\sqrt{3}$. The triangle is equilateral. **2:** If a is the altitude of the cone, the height of the cylinder is $a/3$ and the diameter of its base is $2a$. **3:** $y = \sqrt{\left(\frac{2}{3}\sqrt{3}\right)}$. **4:** a, $2a$, $4a$. **5:** $x + x^2/a + x^3/a^2 + \ldots + x^n/a(n-1)$. **7:** 273/11 seconds. **8:** breadth a: $a+1$ if tank is open; $2a$: $a+1$ if tank has lid. **10:** 8.9 mph. **11:** $h = 2r$. **12:** 6 mph. **13:** 0.00033 second. **14:** 5.33 cubic feet.

Glossary of Strange and Exotic Terms

Measurements, Monetary Units, and
Culturally Relevant Words

Descriptions and explanations of terms that may be unfamiliar to the modern reader are listed below. These terms are grouped by the civilizations within which they were used and thus follow the order of the chapters in the text. In a classroom situation, they add an extra learning dimension by requiring students to research the meaning of unfamiliar terms used in historical mathematics problems. Be aware that over historical periods, meanings and values for many terms have varied and have differed by culture. For example, a yard in the English system is 0.9144 meter or 3 feet, the average pace of a European man, whereas in traditional China, a yard or *bu* was the double pace of a Chinese man, or 5 Chinese feet, *chi*. Where possible, modern equivalents have been provided for units of measurement, but these are approximations.

Babylonia

ban: measure of capacity; 10 L
cable: measure of area; 360 m^2
gin: measure of area; 1 m^2
gin: measure of weight; $\frac{1}{60}$ of a manna
gur: measure of capacity; 3000 L or 0.3 m^2
kus: Babylonian cubit; 0.497 m
manna: basic unit of weight; $\frac{1}{2}$ kg
mina: a coin. See *talent*
sila: measure of capacity; 1 L
shekel: a coin. See *talent*
talent: basic monetary unit; 1 talent = 60 mina = 60 shekels
sar: measure of area, usually the size of a garden plot; 36 m^2

Ancient Egypt

cubit: basic unit of linear measure; common cubit = 0.4 m, royal cubit = 0.525 m

hekat: capacity measure for grain or bread, varying by dynasty; generally ⅟₃₀ of a royal cubit; 4.782 L

pesu: measure of brewing strength for beer

seqt or **seket:** measure of inclination (horizontal run / vertical rise); equivalent to the modern cotangent function in trigonometry

spelt: primitive variety of wheat

Ancient Greece

mina, plural **minae:** a silver coin. See *talent,* below

talent: measure of silver determining monetary system values; 1 talent = 60 minae = 6000 drachmae

Ancient China

bu: measure of length equal to a double pace, or 5 Chinese feet, chi

dan: measure of weight; 60 kg

dou: measure of capacity; 10 L

hu: measure of capacity; 1 hu = 10 dou = 100 L

li: Chinese "mile"; 1 li = 180 zhang = 360 bu = 1500 chi

mu: measure of area; 1 mu = 240 bu²

India

angula: unit of linear measure; a finger width, 1⅜ inch

hasta: measure of length; 1 hasta = 24 angulas = 45 cm

pala: measure of quantity

panas: a copper coin

yojana: measure of distance; varied from 5 to 10 km

Islam

dirham: monetary unit. See *dinar,* below

dinar, plural **denarii:** a gold coin struck in the early days of Islam. The name is derived from the Roman *denarius,* a coin initially valued as the price of ten asses. Due to its recognizable value, it became a standard currency in Europe between the tenth and twelfth centuries.

Medieval Europe

bezant: popular name given to any gold coin—for example, a dinar

leuca: a league; the distance a person could walk in 1 hour, 1.5 Roman miles or 2 km

metreta, plural **metretae:** measure of capacity; 1 metreta = 72 sextarii = 3.4 L

modia: a grain measure

palm: measure of length, the width of four fingers; 7.6 cm

perch: a surveyor's measure of length; 5.029 m

pound: the basic monetary unit established by Charlemagne. A pound of pure silver was struck into 240 pennies; one penny = 1.7 g. The existing coin, the soldi, then became valued at 20 pennies. Thus, a standard was established: 1 pound = 20 soldi or shillings (s) = 240 pennies (d).

sextarius: see *metreta*, above

Renaissance Europe

bezant: see above, under Medieval Europe

braccio, plural **braccia:** Italian measurement determined by the length of an arm; 66–68 cm

dinar, denarii: see above, under Islam

ducat: a gold coin minted in Venice in 1284; 1 ducat = 3.59 gm

florin: gold coin minted in Florence, 1252–1525; 1 florin (fl) = 3.5 g

grossi: a Venetian coin; 24 grossi = 1 ducat

guilden: Dutch florin

leuca: see above, under Medieval Europe

lire: monetary unit; the Italian pound

livre: monetary unit; the French pound, used until 1795

scudo, plural **scudi:** a large silver coin first minted in Milan in 1551. Eventually, the word *scudo* became a popular name for all silver coins.

Japanese Temple Problems

hiki: monetary unit; 1 hiki = 10 *mon*

ken: measure of length; 1.82 m

koku: measure of capacity; 20 L

mon: monetary unit. See *hiki*, above

ri: Japanese "mile"; 414 m

sho: measure of capacity; 0.2 L

sun: measure of length; 2.3 cm

syaku: measure of length; 1 syaku = 10 sun = 23 cm

tan: measure of area; 991.7 m^2

yen: modern monetary unit

England and Colonial America

cathetus: the side of a right triangle perpendicular to the base

crown: British silver coin; 1 crown = 5 shillings

farthing: antiquated British monetary unit; 1 farthing = ¼ penny

guinea: a gold coin minted in England, 1666–1813; 1 guinea = 1 pound 1 shilling

hogshead: a large cask or barrel with a capacity of 63–140 gallons

pound: basic monetary unit of England; 1 pound (£) = 20 shillings (s) = 240 pennies (d)

quintal: unit of weight; 1 quintal = 100 kilo

Other

cavan: rice measure instituted in the Philippines by nineteenth-century Spanish occupiers; 1 cavan = 75 L

Gausthause (modern German, Gasthaus): German guesthouse or inn

halberdiers: bearers of halberds, a combination battleaxe and pike mounted on a 6-foot staff and a popular European weapon of the fifteenth and sixteenth centuries

Reichsmark (RM): currency in use in Germany from 1924 to 1948

rubii: a measure of capacity used for grain

sagitta: Latin for "arrow," as in an arrow on a bow. In a mathematical context, a *sagitta* is the altitude of the segment of a circle.

Bibliography

Works marked by an asterisk (*) are rich resources for extended problem selections for the civilizations surveyed in this book.

Adams, D. 1821. *The scholar's arithmetic; or, The federal accountant.* Keene, NH: John Prentiss.

Ahmes, ca. 1650 BCE. Rhind Mathematical Papyrus. Translated in Chace 1979.

Alcuin of York. ca. 800. *Propositiones ad acuendos juvenes.* Translated in Hadley and Singmaster 1992.

Ascher, Marcia. 1991. *Ethnomathematics.* Pacific Grove, CA: Brooks/Cole Publishing.

Baker, H. 1562. *The well-spring of sciences.* London.

Ball, W. 1987. *Mathematical recreations and essays.* New York: Dover.

Barrow, I. 1665. *Euclid's "Elements."* London.

Bonsall, M. 1905. *Primary arithmetic.* Manila: World Book Co.

Borghi, P. 1484. *Oui comenza la nobel opera de arithmetia.* . . . Venice.

Brooks, E. 1863. *The normal written arithmetic.* Philadelphia: Sower, Potts & Co.

———. 1873. *The normal elementary algebra.* Philadelphia: Christopher Sower Co.

Buteo, J. 1559. *Logistica quae & arithetica vulgò in libros quinque digesta.* Lyon.

Calandri, F. 1491. *Arithmetic.* Florence.

Cardano, G. 1539. *Practica arithmetica et mensurandi singularis.*

*Chace, A. B. 1979. *The Rhind mathematical papyrus.* Washington, DC: National Council of Teachers of Mathematics.

Clavius, C. 1583. *Arithmetica prattica.* Rome.

*Colebrooke, H. T. 1817. *Algebra, with arithmetic and mensuration from the Sanscrit of Brahmegupta and Bhascara.* London: John Murray.

*Cooke, R. 1997. *The history of mathematics: A brief course.* New York: John Wiley & Sons.

Diggs, T. 1572. *An arithmeticall militare treatise named Stratioticas.* London.

Douglas, W. 1814. Question 2. *The Analyst*, March 1, p. 21.

*Eves, H. 1990. *An introduction to the history of mathematics.* Philadelphia: Saunders College Publishing.

*Fukagawa, H., and D. Pedoe. 1989. *Japanese temple geometry problems.* Winnipeg, Canada: Charles Babbage Research Centre.

*Fukagawa, H., and D. Rigby. 2002. *Traditional Japanese mathematics problems of the 18th and 19th centuries.* Singapore: Science Culture Technology Press.

*Fukagawa, H., and T. Rothman. 2008. *Sacred mathematics: Japanese temple geometry.* Princeton: Princeton University Press.

Ghaligai, F. 1521. *Practica d'arithmetica.* Florence.

*Hadley, J., and D. Singmaster. 1992. Problems to sharpen the young. *Mathematical Gazette* 76 (475): 102–26.

*Heilbron, J. L. 2000. *Geometry civilized.* Oxford: Oxford University Press.

Hermelink, H. 1976. Arabic recreational mathematics as a mirror of age-old cultural relations between eastern and western civilizations. In *Proceedings of the First International Symposium for the History of Arabic Science*, April 5–12, 1976, 44–92. Aleppo: Aleppo University.

Hodder, J. 1683. *Hodder's arithmetick.* London: The Rose and Crown.

Hogan, E. R. 1977. George Baron and the mathematical correspondent. *Historia Mathematica* 4:157–72.

Høyrup, J. 1985. Varieties of mathematical discourse in pre-modern socio-cultural contexts: Mesopotamia, Greece and the Latin Middle Ages. *Science and Society* 69 (1): 4–41.

Jiuzhang suanshu. ca. 100. Translated in Shen, Crossley, and Lun 1999.

*Katz, V. 2003. *A history of mathematics.* New York: Addison-Wesley.

*———, ed. 2007. *The mathematics of Egypt, Mesopotamia, China, India and Islam: A sourcebook.* Princeton: Princeton University Press.

Köbel, J. 1514. *Rechenbuck aff linen und ziffern.* Augsburg.

Ladies Diary. 1704–1841. London. Magazine for women published annually.

*Lam, L. Y., and T. S. Ang. 1992. *Tracing the conception of arithmetic and algebra in ancient China: Fleeting footsteps.* Singapore: World Scientific.

*Lenchner, G. 1983. *Creative problem solving in school mathematics.* Boston: Houghton Mifflin.

Leonardo Pisano [Fibonacci]. 1202. *Liber Abaci* [The book of computation]. Translated in Sigler 2002.

*Libbrecht, U. 1973. *Chinese mathematics in the thirteenth century: The Shu-shu chiu-chang of Ch'in Chiu-shao.* Cambridge: MIT Press.

Lieske, Spencer. 1985. Right Triangles II. *Mathematics Teacher* 78:498–99.

Li Zhi. 1248. *Ceyuan haijing* [Sea mirror of circle measurement]. Beijing.

Loci (previously *Convergence*). Mathematical Association of America (MAA) Mathematical Sciences Digital Library website, at http://mathdl.maa.org/mathDL. Washington, DC: Mathematical Association of America. An e-journal of the Mathematical Association of America.

Martzloff, J. C. 1997. *A history of Chinese mathematics*. New York: Springer-Verlag.

Melville, D. 2004. Poles and walls in Mesopotamia and Egypt. *Historia Mathematica* 31:148–62.

Metrodorus. ca. 500. Palatine Collection, or Greek Anthology. Translated in Page et al. 1916.

Milne, W. 1892. *Standard arithmetic*. New York: American Book Co.

Nahin, P. J. 2007. *Chases and escapes: The mathematics of pursuit and evasion.* Princeton: Princeton University Press.

Nemet-Nejat, K. R. 1988. Cuneiform mathematical texts as training for scribal professions. In *A scientific humanist: Studies in memory of Abraham Sachs*, ed. E. Leichty, 285–300. Philadelphia: University of Pennsylvania Press.

———, 1993. *Cuneiform mathematical texts as a reflection of everyday life in Mesopotamia.* New Haven, CT: American Oriental Society.

*Neugebauer, O., and A. Sachs. 1945. *Mathematical cuneiform texts*. New Haven, CT: American Oriental Society.

New York Times. 1969. China's new math and old problems. March 9, p. 18.

Nissen, H. J., P. Damerow, and R. K. Englund. 1993. *Archaic bookkeeping: Early writing and techniques of economic administration in the ancient Near East.* Chicago: University of Chicago Press.

Page, T. E., et al., eds. 1916. *The Greek anthology.* Loeb Classical Library, vol. 5. Cambridge: Harvard University Press.

Perl, T. 1979. *The Ladies Diary* or the *Woman's Almanack*, 1704–1841. *Historia Mathematica* 6:36–53.

Pike, N. 1788. *A new and complete system of arithmetic composed for the use of citizens of the United States.* Newbury-port, MA.

Pine, L. 1997. Nazism in the classroom. *History Today* 47:22–27.

*Plofker, K. 2009. *Mathematics in India.* Princeton: Princeton University Press.

Polya, G. 1971. *How to solve it: A new aspect of mathematical method.* Princeton: Princeton University Press.

Powell, M. A. 1988. Evidence for agriculture and waterworks in Babylonian mathematical texts. *Bulletin on Sumerian Agriculture* 4:161–72.

Pressman, I., and D. Singmaster. 1989. The jealous husbands and the missionaries and cannibals. *Mathematical Gazette* 73:73–81.

Rebstock, U. 2007. Mathematics in the service of the Islamic community. Paper presented at the Fifth European Summer University on History and

Epistemology in Mathematics Education, July 19–24, Prague, Czech Republic.

Recorde, R. 1551. *Pathway to knowledge.* London.

———, 1557. *The whetstone of witte.* London.

Riese, A. 1522. *Rechnung auff der linien und federn.* Erfurt.

*Robson, E. 2007. Mesopotamian mathematics. In *The mathematics of Egypt, Mesopotamia, China, India and Islam,* ed. V. Katz, 58–181. Princeton: Princeton University Press.

Rudollf, C. 1526. *Kunstliche rechnung mit der ziffern und mit den zal pfenninge.* Vienna.

*Sanford, V. 1927. *The history and significance of certain standard problems in algebra.* New York: Teachers College, Columbia University.

Schmandt-Besserat, D. 1992. *Before writing: From counting to cuneiform.* Austin: University of Texas Press.

*Shen, K. S., J. N. Crossley, and A. Lun. 1999. *The nine chapters on the mathematical art.* Beijing: Science Press.

Sigler, L. E. 2002. *Fibonacci's Liber Abaci.* New York: Springer.

*Simonson, S. 2000. The missing problems of Gersonides: A critical edition. *Historia Mathematica* 27:243–302 (pt. I), 384–431 (pt. II).

Smith, D. E. 1917. On the origin of certain typical problems. *American Mathematical Monthly* 24:64–71.

———, 1918. Mathematical problems in relation to the history of economics and commerce. *American Mathematical Monthly* 25:221–23.

———, 1958. *History of mathematics.* 2 vols. New York: Dover.

Smith, J. H. 1880. *A treatise on arithmetic.* London: Rivingtons.

Sun Zi. ca. 400. *Sun Zi suanjing* [Master Sun's manual]. Translated in Lam and Ang 1992.

*Swetz, F. J. 1987. *Capitalism and arithmetic: The new math of the 15th century.* Chicago: Open Court.

———, 1989. An historical example of mathematical modeling: The trajectory of a cannonball. *International Journal of Mathematical Education in Science and Technology* 20:731–41.

———, 1993. Back to the present: Ruminations on an old arithmetic text. *Mathematics Teacher* 86:491–96.

*———, 1994. *Learning activities from the history of mathematics.* Portland, ME: J. Weston Walch.

———, 1995. To know and to teach: Mathematical pedagogy from an historical context. *Educational Studies in Mathematics* 29:73–88.

———, 1996. Enigmas of Chinese mathematics. In *Vita Mathematica,* ed. R. Calinger, 87–97. Washington, DC: Mathematical Association of America.

————, 1999. Mathematics for social change: United States experience in the Philippines, 1898–1925. *Bulletin of the American Historical Collection Foundation* 27:61–80.

*Swetz, F. J., and T. I. Kao. 1977. *Was Pythagoras Chinese?* University Park, PA: Pennsylvania State University Press.

Ticknor, A. 1845. *The Columbian calculator.* Philadelphia: Kay & Troutman.

Trenchant, I. 1566. *L'Arithmetique de Ian Trenchant, departie en trois livres.* Lyon.

————, 1578. *L'Arithmetique.* Lyon.

Treviso arithmetic [Arte dell'abbaco]. 1478. Treviso. Translated in Swetz 1987.

*Van der Waerden, B. L. 1983. *Geometry and algebra in ancient civilizations.* New York: Springer.

Verschaffel, L., B. Greer, and W. Van Dooren, eds. 2009. *Words and worlds: Modeling verbal descriptions of situations.* Rotterdam: Sense Publishers.

Vogel, K. 1983. Ein Vermessungsproblem Reist von China nach Paris. *Historia Mathematica* 10:360–67.

Watson, E. 1777. *Connecticut Almanack.* Hartford, CT: Eben Watson.

Wikenfield, M. 1985. Right triangle relationships. *Mathematics Teacher* 78:12.

Wittfogel, K. A. 1957. *Oriental despotism: A comparative study of total power.* New Haven, CT: Yale University Press.

Yung-lo ta-tien [Yung-lo encyclopaedia]. 1407. Photographic repr., 1960. Hong Kong: Chung-hwa Book Co.

Index